GW01117253

POWER ON THE LAND
A Career in Agricultural Engineering

My Life and My Work Series

Accountancy *Taste for Accounting* by Peter Reynolds
Advertising *One Off* by T. E. Johnson
Agricultural Engineering *Power on the Land* by Brian May
Armed Forces: Navy, Army & Air Force *All the Queen's Men*
Art: Painter & Teacher *A Painter's Diary* by Francis Hoyland
Author & Writer *The Ring of Words* by D. J. Hall
Banking *Wealth of Interest* by J. B. Pickerill
Child Care *Children in Jeopardy* by Joan Lawson
Dentistry *Chew This Over* by Matthew Finch
Divine Ministry: Anglican, Roman Catholic & Methodist *Rev*
Education Services *Beyond the Classroom* by J. M. Hogan
Electrical Engineering *Making Electric Things Happen* by P. L. Taylor
Exporting *Exportsmanship* by Hamish Lindsay
Farming *Down to Earth* by Donald Knight
Hospital & Community *Medical Social Work* by Helen Anthony
Hotel & Catering Industry *At Your Service* by Julian Morel
Housing Management *Real Estates* by Prudence Baker
Journalism *Dateline: Fleet Street* by Owen Summers and Unity Hall
Law *Let's Make it Legal* by John Malcolm
Librarianship *Books are for People* by James Dearden
Marketing *A Consuming Passion* by Martyn Bittleston
Mechanical Engineering *Of Men and Machines* by George Stevenson
Medical Laboratory Technology *Ask The Lab* by Kenneth Hughes
Medicine *I Swear by Apollo* by P. T. Regester
Merchant Navy *The Dustless Road* by S. J. Harland
Mining Engineering *Mines and Men* by Tom Ellis
Personnel Management *Man Power* by John Brandis
Physiotherapy *Towards Recovery* by Pat Waddington
Prison Service *Prison People* by Nicholas Tyndall
Public Health Service *The Environmentalists* by Peter Brooks
Publishing & Bookselling *Books are my Business* by A. W. Reed
Radio & Television *The Informing Image* by Rodney Bennett
Retailing *Can I Help You Madam?* by Marian Aitken
Scientific Research *Why Research?* by Sir Ian Wark
Secretarial Skills *Take a Letter Please* by Mona Tratsart
Speech Therapy *Speak for Yourself* by Betty Brown
Statistician *My Statistics are Vital* by Madge Dugdale
Teaching *And Gladly Teach* by Margaret Miles
Theatre Management *Beginners Please* by Elizabeth Sweeting
Women Police *The Gentle Arm of the Law* by Jennifer Hilton
WRNS *Never at Sea* by Vonla McBride

My Life and My Work Series

POWER ON THE LAND
agricultural engineering

by Brian May
BSc CEng MIMechE NDAgrE MIAgrE

with a foreword by Professor P. C. J. Payne
MSc PhD MIAgrE MASAE

Educational Explorers · Reading

First published 1972
© *Brian May* 1972
isbn 0 85225 740 6

Published by Educational Explorers Limited
40 Silver Street, Reading, England
Set in 'Monotype' Bembo and printed in Great Britain by
Hazell Watson & Viney Ltd, Aylesbury, Bucks

CONTENTS

Page

Foreword by Professor P. C. J. Payne 9

1. Why Agricultural Engineering? 13
2. On the Farm 21
3. 'Voluntary' Service Overseas 30
4. Practical Experience in Horticulture 39
5. Back on Course: The National Diploma in Agricultural Engineering 43
6. Design Trainee with Massey-Ferguson: Degree Studies in Mechanical Engineering 60
7. Wider Industrial Experience: Production and Marketing Training 71
8. Professional Status in Sight: Applied Research .. 82
9. Making Farm Machines Work: as a Design Engineer 90
10. The Other Side of the Bench: Educating the Agricultural Engineers of the Future 97
11. Accepting the Challenge: Research and Development Projects 105
12. The Proof of the Pudding 114
13. If I could Start Again 122

Appendix: Some Helpful Information 133

ACKNOWLEDGEMENT

The Author and the Publisher gratefully acknowledge the kind help and advice given by

THE INSTITUTION OF AGRICULTURAL ENGINEERS

MINISTRY OF AGRICULTURE, FOOD AND CO-OPERATIVES, TANZANIA

ALASDAIR BARRON

JIM KELLAWAY

DAVID MORRISH

TERRY PODMORE

MARTYN RIDDLE

and are indebted to the following for permission to publish the illustrations in this book

NATIONAL COLLEGE OF AGRICULTURAL ENGINEERING

NATIONAL INSTITUTE OF AGRICULTURAL ENGINEERING

BOB MANN
Projects Officer, Intermediate Technology Development Group

MASSEY-FERGUSON (UNITED KINGDOM) LIMITED

RANSOMES, SIMS AND JEFFERIES LIMITED

GORDON SPOOR

BRUCE WITHERS

ILLUSTRATIONS

AGRICULTURAL ENGINEERS OF THE FUTURE
Experiment with a model tractor at Silsoe 1
Students testing a mechanical potato planter .. 2

AGRICULTURAL ENGINEERING: A BROAD SUBJECT
Design: a blackcurrant harvester 3
Control of the Environment: a mist-propagator and
heating stimulates growth in a seedling nursery .. 4
Field Engineering: stereoscopic plotter aids
preparation of irrigation and drainage schemes .. 5
Research: study of tractor cab noise levels 6

MANUFACTURE OF FARM MACHINERY
At Home ... combine harvester assembly line .. 7
Final stage in farm tractor assembly 8
And Overseas ... a simple mechanical cultivator .. 9
A bicycle-operated separator for cereal crops .. 10

MECHANIZATION
Ploughing heavy land in East Anglia 11
Cultivating rice paddies in Ceylon 12
Starting work on a farm in Japan 13
Harvesting sugar cane in Jamaica 14

IRRIGATION AND LAND DRAINAGE
An Iraqi irrigation project 15
Laying tile drains to lower the water table 16

Inset at pages 48 to 57

FOREWORD

by

PROFESSOR P. C. J. PAYNE

M.Sc., Ph.D., F.I.Agr.E., M.A.S.A.E.

IN MY OPINION, 'The Quality of Life' could prove to be the most important phrase that has come into the English language in this decade. Of course, the words themselves are only a new way of expressing an unquenchable idea. We used to call it 'Human Happiness' and the like. The important thing is that this generation has manifested its re-dedication to the old pursuit by the coining of a catchy phrase; and that this generation, more than any before, possesses the power to make a real improvement in the quality of its own life and that of its successors.

Almost all occupations, if pursued ethically and vigorously, provide some opportunity to improve the quality of life and it is particularly easy to point to the ways in which Agricultural Engineering, the one described in these pages by Brian May, makes its contribution.

Firstly, it provides the tools for the world's most important industry—farming—which occupies more people producing a more vital consumable than any other. It is a strange paradox that those parts of the world with the most intensive manufacturing industry also contain the most productive farming. The value (more than £2,000,000,000) of British agricultural output at least equals that both of Canada and of Australia and far exceeds that of most other primary producing countries.

The productivity of British farming has risen at double the national average for all industries since the war and one of the main reasons is that competition for labour leads to the more intensive application of the varied techniques of the engineer. This high demand for engineering skills and products in turn raises their quality and, in the case of products, lowers the price. So a career in agricultural engineering can provide a sense of belonging to something large, competitive, stable and of real value to mankind.

Secondly, agricultural engineers from the richer countries possess in their know-how and their products one of the most sought-after commodities by the poorer nations who must improve their agriculture to create the spending power, to justify the industries that eventually can narrow the wealth gap. So agricultural engineering holds one of the keys to bringing the two halves of the world together.

Thirdly, the quality of life depends, besides the creation of wealth, upon the environment in which we live and play and since agricultural engineering practitioners possess great power both to improve and to devastate the countryside, it is vital that they should exercise their power with wisdom. In the past they have played a major role both in maintaining rural order and beauty against apparently crippling depopulation and in creating deserts out of gardens.

It was soil erosion caused by the excesses of irrigation engineers that started the downfall of the civilisation of Mesopotamia three millenia ago and it was the misuse of relatively modern cultivation equipment that caused the dust-bowl of North America in much more recent times. The need for environmental awareness is one of several reasons why the fascinating techniques of modern Land Resource Planning are becoming more and more the concern of agricultural engineers and why their subject can be regarded as an applied environmental science.

By now, the reader is probably wondering if there is any facet

FOREWORD

of life not touched upon by agricultural engineering, which brings me to my fourth point—the spice brought to life by the variety of its interest. It ranges from soil conservation in Malawi through refrigerated fruit stores in Maidstone and tractors in Melbourne to the prevention of back-ache for hoers in Middle Wallop. As a technical profession it is a branch of engineering and as a sphere of activity it is a part of farming.

But in spite of all the fine words and aspirations, the quality of life for an individual depends greatly upon what he does in everyday life, so read on.

PETER PAYNE

Clophill,
Bedfordshire.

I

WHY AGRICULTURAL ENGINEERING?

THE NEW SESSION STARTED almost a week ago. Students and many members of staff are gathered in the Junior Common Room enjoying a drink and chat before the Fresher's Dinner. This occasion is a regular feature in the social calendar of the College and provides one of several opportunities during the year for students and staff to meet outside the academic programme.

The College, or to give it its full name, The National College of Agricultural Engineering, was set up almost ten years ago to meet the needs of a young and rapidly growing industry. Student numbers have increased from fifteen to nearly two hundred in the ten year period. A quick glance around the Junior Common Room indicates that although numbers are small, a wide range of nationalities and ages are represented.

I am talking with a group of students standing near the bar, in a secluded part of the Common Room called 'The Joint', an area in which staff and students can meet at any time. Our conversation is being made against a background of constant indecipherable chatter resulting from many groups similar to ours talking and laughing at the same time—a real sign that the objectives of the occasion are being achieved.

As the conversation proceeds I wonder just how many of our new arrivals know why they are here. Perhaps an unfair question since they have not yet had a chance to sample the course of study.

Are they here by chance or by choice? Do they have any idea of what agricultural engineering is really about? I wonder how many students are starting courses elsewhere who would really be much better suited to agricultural engineering?

Are there some students halfway through their course who would give their right arm to be studying elsewhere? What about the thousands of young people who are now in their final year at school? Will they consider agricultural engineering, and if so what facts and opinions will they have available when making this consideration?

Choosing a career is an extremely difficult and chancy business for most students. Circumstances, opportunities, influences from home, school and friends all play their part.

Richard's father is a distributer of farm machinery. His choice was easy once he knew that there were prospects for him in the family firm. A three year B.Sc. course in agricultural engineering should be an excellent preparation for the future career which lies before him.

It wasn't nearly so easy for Ann who is now in the second half of the four year honours undergraduate course. Her father is a solicitor which is a long way removed from agricultural engineering! She lives in a large town in the Midlands, seeing very little of agriculture and having no contact with engineering. One summer towards the end of her school career, her family rented a country cottage for two weeks. Each member of the family had a share in the running of the cottage. Ann's job was to collect the milk from the nearby farm each morning.

She was interested in what she saw of farm life and when the time came to choose a career this experience was taken into account. The number of girls studying engineering is increasing each year and Ann decided to join the ranks.

During last summer vacation Ann spent two months with a London based firm of consulting engineers assisting with report preparation on overseas rural development projects. She would like to carry on with this work full-time after graduation.

Richard and Ann are examples of students who have managed

to find their way early on. For many, the path is not nearly so straightforward. Jim is now in his third year and still does not know what he wants to do on leaving College.

I can remember meeting him for the first time at his College interview and have met him on several occasions since for tutorial work. He is the sort of young man who can work much more enthusiastically if he can see how theory is applied in practice. This has led us into several discussions about the subject of agricultural engineering from designing tractors to making canals for irrigation. Although he has demonstrated a useful design ability he is not sure whether he would choose to be doing this all the time. I hope that for him as well as many others, practical experience gained during vacations combined with discussions and coursework at College will help to clear the mist before his course is completed.

The Fresher's Dinner provides a good opportunity for me to get some answers to my questions. I found myself sitting at a table in the Dining Hall next to David and Phil who had just arrived at College.

The walls of the Dining Hall are decorated with shields on which are painted the national flags of students who have attended courses at the College. This always provides a good opening to conversation—especially as they number over forty. Since the College began, close attention has been paid to the special problems of overseas agriculture particularly in the developing countries. This has had the effect of attracting many students from overseas mainly for postgraduate courses.

David enquired whether any opportunities existed for British graduates to work overseas. His reason for raising this point was because he wanted to do a job which involved travel to other countries.

I was able to assure him that several of our graduates are now working in countries all over the world in a wide variety of jobs. Although most developing countries are now independent, it is likely that for several decades they will still wish to import technologists—particularly agricultural engineers. Common Market developments and the need to increase exports will also mean that several engineers working for companies in the United Kingdom will be required to make overseas visits.

David had just spent two years on a National Certificate course in Mechanical Engineering but with emphasis on agricultural engineering. His good marks in the final exams had enabled him to get on to the degree course. Between five and ten per cent of our students enter the course in this way each year.

He had managed to get two 'O' levels at his local comprehensive. His form master, who was also responsible for advising on careers, suggested a Technical College course as being 'best suited to his abilities'. Since he had been brought up on the family farm in the West Country he wanted to apply engineering to agriculture, hence the O.N.D. with an agricultural engineering bias.

Clearly by coming on to a degree course, David's choice of job will be wider later on and this will probably be reflected in a higher salary. But, as I so often point out to young people, a qualification is not a passport to success. It is a 'visa' or entry permit only. Personal qualities are equally important. In later years the significance of initial training and qualification recedes while personal qualities and experience increase in value as greater responsibility emerges.

Both David and Phil accepted this point which I have also found valuable and constructive when talking to the inevitable but small number of students who fail the course at some stage.

The question of salary prospects naturally arises when discussing careers and it was Phil who made the point. He readily appreciated that it was impossible to give specific figures especially as they seem to rise so quickly nowadays.

I was able to assure him that salary levels for agricultural engineers are at least equal to those in other engineering professions. In some job areas salaries are higher but this of course again depends on level of responsibility, personal qualities and experience.

It is quite wrong in my opinion to rate a job by salary alone. After all, most people spend about forty-five years working, so job satisfaction must be high on the list, followed rather than preceded by salary.

Phil's most interesting contribution to the dinner discussion was concerned with specialisation. He had been advised by his Careers Officer to take a general degree in Engineering at a large University. The course at Silsoe was considered to be a somewhat specialised branch of engineering which would be better studied at higher degree level.

It was an agricultural engineering friend of the family who finally influenced Phil but there was still some doubt in his mind.

I readily admitted that there were several points of view concerning the choice of College or University and the course of study. Naturally, I am biased because I believe that agricultural engineering and those associated with the profession have a great future in the modern world.

It is a broad subject with many aspects. These include agricultural, social and economic as well as engineering. Then there are many sides to the engineering aspect. Agricultural engineers need to know something about civil, mechanical and electrical engineering. Even architectural considerations are relevant to some work. I was therefore unable to accept that agricultural engineering is particularly specialised. Relative to many subject studied it is broad, challenging and very relevant to the needs of modern society all over the world.

Phil seemed to be satisfied with these remarks which I hoped would strengthen his convictions about his recent decision.

After dinner the evening continued in the Junior Common Room with further conversation and alcoholic beverage. The topics had by now shifted to more domestic matters such as the likely success of sports teams and effects of delays in the latest building programme.

Later that evening I found my thoughts returning to the original questions but this time in relation to my own experiences in my 'early days'.

Just how did I develop this interest in agricultural engineering? It must have started during my school days because immediately on leaving school I was preparing for entry to a course leading to the National Diploma in Agricultural Engineering.

Living in small villages during the whole of my school career had an important influence on my attitudes towards those subjects which I found to be of interest. It was frequently possible to relate science subjects to rural life. In this respect I had a distinct advantage over my urban friends.

For me, physics became the physics of farming. My major source of pocket money was weekend and vacation work on a 1,500-acre farm surrounding my home. The knowledge and experience gained working on the land, occasionally driving, operating and in some cases taking machinery to pieces was an important influence I am sure—although probably I didn't realise this at the time. The work that went on in the farm mechanic's well-equipped workshop always interested me. I can remember the fascination of watching a heap of nuts, bolts and various pieces of metal being put together to form a re-conditioned tractor engine and then watching the assembly spring to life when supplied with fuel.

Of course there were phases of rebellion and rejection—periods which must have tried the tolerance and patience of my parents to the extreme. My lack of attention to study and not infrequent dismal school reports must have on occasion driven them near to despair. Their efforts were not appreciated at the time but I can now recognise the value of their background support and guidance suspecting that even to this day I am grossly underestimating its value.

My main problem at school was recognising the fact that the world does not revolve around sport. There was a time when I saw myself as a member of the county cricket team and a professional football player during the winter months. I hadn't quite decided how to fit in the athletics season and a keen interest in hockey when the postman delivered my all too short list of 'O' level results.

The headmaster of my grammar school spoke to me in a firm but friendly tone pointing out the value of qualifications which up to that time I had never accepted as particularly important. He suggested that I stay on at school for at least another year to improve my 'O' level score and do some 'A' level work. I was all for leaving school at that moment, doing well in the County Trials to be held on what was to me the sacred turf of Canterbury cricket ground and then joining the ground staff in preparation for a rapid rise to the first team and then on to Test Matches!

My headmaster had other ideas however, and the Trials in which I took part the following summer made me realise that he was right. Four runs scored and no wickets for thirty-eight runs in four overs against schoolboys on a bowler's wicket is hardly Test Match standard! How glad I was of the first year in the VIth form. My interest in academic work was increasing all the time and being a prefect I also gained my first experience in the use of authority and acceptance of responsibility. This latter comment is not meant to imply that I was previously irresponsible—but I certainly began to look at things from a different angle!

Failure at 'O' level gave me a jolt. I took some subjects again at the end of my first year in the VIth form to give an air of respectability to the final list.

I shall never forget my headmaster's comments during the interview at the end of this year. My internal examination results suffered from the excessive committment to sport. Realising that another year in the VIth would achieve little except perhaps another book token at Prize Day for the '*best contribution to the social and sporting life of the school*' I suggested to the Headmaster that I leave now rather than wait for the inevitable 'A' level failures. He expressed no surprise at this and reacted as though he was about to say the same thing himself which did my ego no good at all!

'You would do very well to think about engineering in agriculture' he said, 'we must always learn effectively to combine whatever talents are present within a person. With your signs of ability in the sciences, undoubted sporting interests and not unpleasant rural accent I think—yes I am sure—that you would instil confidence in the farming fraternity. If I were wanting to buy a tractor I would be quite likely to buy one from a fellow such as yourself'.

Instead of becoming a tractor salesman, which didn't appeal to me at the time, I started to think about further education.

I had heard about the National Diploma in Agricultural Engineering from a farming magazine and thought this would be worth trying for. It involved spending one year on a large mechanised farm which could be arranged locally—and for which I would be paid! I would also have to get an Ordinary National Diploma in Mechanical Engineering as an intermediate qualification.

So, on leaving school, I fixed myself up with a one-year apprenticeship on the farm and registered for entry in the agricultural engineering course at the Essex Institute of Agriculture. At the back of my mind I felt that, in addition to having a plan to work to, it was flexible. And if something else turned up, then I was in a position to take anything better.

2

ON THE FARM

CAREERS ADVICE is far easier to give than accept. I have attended several Careers Day activities in Bedfordshire presenting the case for Agricultural Engineering. They have been very well presented with school halls filled with posters, pamphlets and displays. Many classrooms are used for lectures, filmshows and discussion groups. Engineers, doctors, nurses, accountants, soldiers, policemen and a whole host of representatives from other walks of life are there. It is a bewildering sight to me and even allowing for the thoughtful guidance and advice given by masters and mistresses it must be bewildering for many of the young people trying to find a slot into which they think they might fit.

I had only my Headmaster and local Youth Employment Officer to turn to and although the Headmaster's advice was specific, several alternatives were presented by the Youth Employment Officer, who terminated his interview by saying 'It's now up to you, young man, do let me know if you want any further advice. There are other opportunities to consider, of course. If you would like to come back another time, I will discuss them with you'.

Decisions, decisions! Life at school is so well organised. You don't have to decide what to do tomorrow—it's there written down on the timetable.

I was in a turmoil for three weeks after the end of term. Which way should I jump? Two engineering firms looked as though they would accept me for a five-year apprenticeship, marks gained in the Chatham Naval Dockyard apprenticeship entrance examination gave me a choice of five departments, and then there was a whole year on the farm with more full-time study at the end of it.

What made me choose the farm? Was it returning to something that I knew and had enjoyed during school holidays and weekends? Was it the outdoor life? A reluctance to sign on for five years? Two years National Service was due in twelve months' time. I had no choice about signing on there, so I chose freedom and flexibility, while it was still available!

An agricultural engineer needs to know about farming methods and conditions. He needs both knowledge and experience of these things. The best way of getting experience is to work on a farm. I had already spent many vacations and weekends helping with a variety of jobs, so my year as a farm apprentice consisted of filling in the gaps in my experience working among people that I already knew well.

Farming is a splendid life, working in the open air watching the seasons come and go. Provided you understand the local weather conditions and soil, keeping nature on your side as far as possible, the work is rewarding and satisfying. People working on the land have, however, been decreasing in number for many years. New machines are being made and existing ones improved each year to replace men.

During my year of apprenticeship I saw larger and more powerful tractors introduced on to the farm. A very impressive and complex arrangement of machinery and equipment for grading, drying and storing several hundred tons of corn was installed in time for the harvest while the previous year, a large stationary pea viner had been purchased. This still attracted attention from neighbouring farmers as it was the only one in the area.

Helping with the harvest was my first task and what a hectic period this proved to be. Golden yellow and light brown fields of ripe oats, barley and wheat were methodically consumed by three combine harvesters, which in turn fed a team of trailers drawn by tractors crawling patiently alongside the harvesters amidst a cloud of dust.

Once filled, the trailers would be hauled off to the receiving pit of the grain drier where the load was discharged and the field cycle repeated.

It took me a day or two to recognise the important part that the farm manager was playing in these operations. At first it looked as though he was just enjoying himself driving his Land Rover from field to field, chewing grains of corn and looking at the sky. Then off he would go back to the drier inspecting various dials and chatting to the drier operator.

As I got to know more about the details, I could see that this was sophisticated, scientific farming operating according to a plan which was executed on the lines of a military exercise. The farm manager was a retired army major, experienced in organising men and operations. He was well used to command, and the farm workers did not object to his crisp manner of delivering direct instructions. They respected his ability to get the job organised, knowing that he recognised their superior knowledge of the range of machines upon which they were all heavily dependent.

The men were on parade each morning during harvest in the barn which housed milling and grinding machines for preparation of cattle feed. Pay parade was also held here on Friday evenings. There was a dusty and torn plan of the farm layout mounted on the wall which the Major referred to with a broken billiard cue during the ten-minute operations session. The Major was in his element—no doubt occasionally recalling experiences from World War II.

Working in the drying and storage plant made me feel important. In the field I was only allowed to drive the combines under close supervision and the tractor drivers were reluctant to let me have a go for fear that I would allow the combine to discharge grain on to the stubble. That would have certainly drawn unwelcome attention from the Major!

There was always plenty to do checking on the flow rate through the drier, measuring moisture content of grain in and out of the drier and inspecting the augers and conveyors feeding the storage silos—large cylindrical structures made from special concrete blocks secured by metal bands.

Only one man was normally required to operate the plant, but I always worked as an assistant to the regular attendant. Occasionally he would be called away for an hour or so, when I had got used to the system and it was then that I felt important—recognised as being capable of accepting responsibility.

Straw which was discharged on to the ground by the harvesters was subsequently compressed into rectangular-shaped blocks (or bales) and tied up with twine by another tractor-drawn machine called a baler. The baler collects the straw from the ground by revolving tines and conveys it to the compression chamber. A large ram reciprocates in the chambers driven by a propeller shaft from the tractor. The needles and mechanical device for tying knots in twine is ingeniously designed and helped a lot to assure me that the agricultural engineer is definitely not a glorified blacksmith!

I didn't operate the baler. My height and muscles in the shoulders and arms, well developed by many hours hurling a leather-covered ball and swinging a length of willow, were considered well suited to the task of lifting bales with a pitchfork and projecting them on to a trailer. The physical workload imposed by this task was enormous, especially when the bales became damp or the needle and knotting mechanism had been slow off the mark, leading to an extra long bale.

If only the cattle and pigs who subsequently used this material for bedding could have indicated how much they appreciated my efforts it would have helped! As the pile of bales stacked on the trailer grew higher it reminded me of some Sports Day field event at school—no records or prizes though—only the cattle and pigs benefited.

Surely there must be a better way of dealing with a by-product of the harvest than this? Engineering was making a significant contribution to agriculture, but clearly there was still plenty of room for improvement.

As the evenings began to draw in, the three or four inches of stubble left by the combine harvesters was burnt to prepare the land for ploughing. This always serves to remind me that November 5th cannot be far away—a barbarous but enjoyable event always to be remembered because it is my Mother's birthday.

Once the basic principles of ploughing had been understood—which tractor speed to use, how to start and finish, what needs adjusting when and how—I found the job slightly monotonous but satisfying. In Kent the seabirds are never far away and they always appreciate the juicy worms and grubs exposed by the plough. It is a majestic sight to watch the lapwings wheeling gracefully at speed above the plough—more than adequate compensation for having to sit on the tractor chilled, shaken and awkwardly positioned in order to watch the furrow!

Autumn cultivation of the land goes on for as long as soil and weather conditions permit. It is sometimes a great rush to get the autumn crops planted before things get too difficult. During my year on the farm it was possible to drill about 50 acres of winter wheat which the Major thought was a 'pretty good show'.

The arrival of autumn also meant the start of studies once again. With the National Diploma in Agricultural Engineering course in mind, I spent my one-day release attending the nearest Technical College preparing for the second year examinations of the National Certificate in Mechanical Engineering. My eventually respectable 'O' level results enabled me to gain exemption from the first year of the course. It was much easier to see the practical applications of coursework here than at school. Perhaps because of this I found the work more interesting and easier to understand.

The new and different situation that I found myself in helped a lot. More effort was required on my part—to travel the twenty miles to do something which I had actually chosen. Most of the class were engineering apprentices, from Medway towns, one or two years younger than myself, several of whom regarded the release to College as a release from work—any type of work.

Lectures and practical classes went on all day but the group conversation at morning coffee break was frequently concerned with such matters as which film was worth going to in town that afternoon. 'Playing the system' was to be marked present on the class register at 2.00 p.m. and then to slip out for the afternoon showing of the latest adventures of the current screen heroes. Three quarters of the class, including myself, never took part in the game, although at the time I was never quite sure why.

I would like to think it was because I had made up my mind to succeed at last. This thought might have been in my mind but very deep down and uninfluential. The more likely reasons were an interest in the course, gratitude for being given a day off with pay and fear of being found out!

Back at the farm, the growing season was over and apart from those who had animals to tend, many workers were found maintenance jobs while others prepared crops gathered for storage or sale. Following a brief period riddling and sacking potatoes I was sent off to help the farm mechanic in his workshop.

Bill was a real character, a true Cockney and proud of it. He was also a first-class mechanic and proud of that too. His sense of humour and repertoire of jokes and stories of doubtful origin were second to none.

'Git yerself a trade boy' he would advise. 'Might as well be a barra boy dahn the Lane as spend yer days 'anging abaht 'ere for a pittance. Sure, go to that College if yer fink its any good, but do somfing—can't 'ang abaht vese days. My old woman made me work fer 'er staht money as soon as I left school. Look where that's got me—grease monkey fer a bloody rich farmer!'

I spent almost three months of my time with Bill. He took me under his wing, worked me hard but gave me plenty of experience. He did not suffer fools gladly and with a vocabulary heavily biased towards the words not to be found in a dictionary, he made sure that I was in no doubt about the need to work well, quickly and without mistakes.

Perhaps I should have complained to the Major that I was being sworn at, unfairly treated and a hundred and one other things that I might have objected to. Instead I chose to weather the storm, face up to the situation—and Bill. This I now know was the right decision. Bill had a lot to offer and it was in my interest to accept—on his terms.

His careers advice wasn't presented as eloquently as others I had listened to but to me it was amongst the most important. This wasn't a schoolmaster talking or a professional adviser, but an ordinary member of the working public. One who had had real and tough experiences first hand and had fought his way through them, literally on occasions.

He had little outward respect for authority as shown by his signature tune *Auprès de ma blonde* sung to the single phrase 'All coppers are bastards'. I neither approved nor disapproved of this but accepted it as part of him. His opportunities at my time of life were very limited. He was trying his best to get me to see and accept those which were now available to me. For this I was grateful.

Many of the farm machines came to the workshop for overhaul and repair during the winter months. I was able to inspect them at close quarters, see how they worked and what parts suffered most in the field. I also had the job of checking the spares and tools in Bill's van which he used during the harvest and other busy periods as a service aid to all the machines in the field.

Connected to the farm radio control system Bill and his van travelled many miles in the summer from one ailing machine to another replacing belts that had worn, filters that had become choked with dust and bolts that had sheared.

Towards the end of my stay Bill allowed me to overhaul completely the Major's Land Rover engine myself showing me how to grind valves, hone bores, fit oversize pistons and replace bearings. There was a ghastly fate threatened if anything should go wrong, which I am glad to say wasn't carried out because after one or two breathtaking moments the engine finally spluttered to life when reassembled!

Once the danger of heavy frosts was over the farm bricklayer cum plumber electrician was given the job of building a second new cowshed to enable a very old thatched wooden building, which had been used for milking a small part of the dairy herd, to be demolished. I was quite sorry to see the building go because although it was infested with rats they provided sport during the winter evenings when we attempted to limit their numbers with air rifles.

The technique was to creep up to the building about an hour after dark, open the door gently and quickly turn on the lights. Often a dozen rats could be seen scampering everywhere, on the feed troughs, along the vacuum lines and in the swill gutters. Once established in their runs under the concrete base of a building they are very difficult to get rid of. The solution in this case was to remove the building and break up the concrete.

Some work had already been done the previous summer. The footings and a few courses of bricks had been laid. It was my job to act as the handyman's labourer. An engine-driven mixer helped me to keep him supplied with cement. A constant supply of bricks was also required and although my skin had hardened slightly since working on the farm, it soon became raw at the finger tips after I'd handled a few bricks.

As I was to discover later, farm buildings are very much the concern of some agricultural engineers so it was useful and appropriate experience for me to be involved in this building work right up to fitting the equipment and installations once the main structure was completed.

There was some time for me to take part in the Spring cultivation and planting work. As the soil dries out and becomes workable the machinery and tractor shed begin to empty and another season is under way. By now I was regarded as sufficiently competent to be given a tractor and machine to take out alone. I had, at least in some respects, become a member of the team instead of just an onlooker.

Although I didn't spend a set period of time in the dairy, it was possible for me to call in on occasions while working on the annex. I wanted to get to know something about this part of the farm and was impressed by the standards of cleanliness maintained and the ability of the cows to learn to enter the same stall each time for milking. The machinery used in the diary is specialised but vital to the activity. Vacuum operated milking machines and automatic recording equipment means that two men could milk about a hundred cows. The only job that they had to do by hand was to wash the udders and fit the teats connecting the cow to the machine.

In the early summer I was called for a medical examination by the War Office, as I had elected to get National Service out of the way then, rather than delay it until I had completed any further full time education. I was pronounced fit for service and since it looked as though I would be staying with engineering I asked to be placed with the Royal Electrical and Mechanical Engineers.

So it was that after a few more enjoyable weeks on the farm I was called to serve Her Majesty for two years.

3

'VOLUNTARY' SERVICE OVERSEAS

ALMOST EVERY YEAR, one or two of the graduates from the National College go overseas for a year as a 'V.S.O.' They are taking part in a system operated by the British Government in conjunction with overseas countries which provides young people with first hand experience of working and living in developing areas of the world. At the same time it is intended that the people living in those areas should also benefit in some way from the experience.

Agricultural engineering graduates are ideally suited to many of the projects undertaken since they are often concerned with rural development. The pay is not good but the experience more than compensates for this.

It is stretching a point to the limit by regarding my own initial overseas service as voluntary, but in several respects my postings to Kenya and Malaya during National Service gave me similar experiences. Perhaps I should start at the beginning . . .

It was becoming apparent to me at this time that it is not possible to go through life always doing exactly as one pleases. Tasks sometimes have to be tackled that are not appealing. The best method of dealing with these is often to face them squarely and get them out of the way rather than to keep putting them off. My commitment to National Service fell into this category.

To me, this period was going to be a waste of time, something that I would certainly not have chosen to do but it seemed inevitable and (to others) necessary, so I became determined to make the best of it. After six weeks of 6.30 a.m. breakfasts, cross-country runs, marching, scrubbing floors, peeling potatoes and

'VOLUNTARY' SERVICE OVERSEAS   31

more marching, I was considered mentally and physically prepared for a vehicle mechanics course in wooden huts set in splendidly relaxing Somerset countryside. At least this is how recent holiday advertisements describe it although there wasn't much relaxation to be found while I was there!

The course itself was very useful and intensive. Fourteen weeks divided into seven two-week sessions were spent working through vehicle sections from the engine to the ground drive components. Competent civilian instructors gave talks and demonstrations during each session. There was plenty of practical work which included down to earth exercises in logic called C.W.G. tests (car won't go). Sequence charts were provided and a row of twelve engines each of which had been cunningly tampered with by the instructor. This training tied up well with my workshop experience under Bill on the farm. Gaps were filled in; my knowledge and experience were extended.

At home, school and on the farm, people were often just a part of the overall scene, in many cases neither accepted nor rejected—just there. In the Services, for the first time in my life virtually everything and every person around me was new and unfamiliar. I began to take much more notice of individuals and what they stood for. Did they have any particular characteristics which I hadn't met before? How might these characteristics affect their actions and reactions? The 'haves' and 'have nots' were much more clearly defined. There it was—provocatively chalked on the arm for all to see, or decorously perched on the shoulder just waiting to be saluted.

So much of one's work is concerned with people; working with them, under them, for them—sometimes deliberately against them! Learning how to serve, lead, and gain the respect of your fellow beings is not easily acquired, but once attained, even in a small measure, it is an extremely valuable asset to be fully recognised and developed as far as your ability allows.

In civilian life all this goes on as a free enterprise operation. The ability to get on with and influence people is frequently the deciding factor in the promotion stakes. Authority is something to be kept in the background where possible, used intelligently and sparingly. The aims and needs of military life are often quite different and the use of authority predominates.

I enjoyed the power that my single, often whitened but thin stripe provided. I also soon learnt that my elevation to the lowest non-commissioned rank produced nothing in terms of genuine respect unless I was prepared to earn it as an individual with whatever qualities I might possess. The emphasis on authority and abundance of examples of its misuse was an invaluable experience for later life.

There were of course the lighter moments even when training. The sight of our 6 ft 4 in, twenty stone platoon sergeant cycling slowly on to the square at 7.0 a.m. buttocks drooping over the saddle almost touching the rear mudguard never failed to stir us from our somnolent state. We watched the dismounting procedure as an onlooker gazes at the trapeze artist—waiting for a mistake followed by rapid decent and final collapse in a heap on the floor. It never came.

Continuing my theme of trying to make the best of a bad job I put my name down for overseas service. The decision rested between a home posting and an opportunity of carrying on with my studies at a technical college or seeing something of the World. There seemed to be plenty of time for studies later on. Apart from two sheltered weeks near Dijon with a school party and the French master, I'd never been off the Island, in fact only once north of London, and here was the chance really to spread my wings.

The written announcement informed me that I was to be sent to Kenya and would be attached to the Rifle Brigade—an infantry battalion, but one which I was to discover later had well over 100 vehicles of various types which needed maintenance and repair.

After a somewhat eventful journey in an ancient Hermes we eventually staggered thankfully on to firm ground at Nairobi airport. The overheated engine which I am sure actually caught fire was fortunately still intact. We were drenched in a downpour at Naples airport, almost eaten alive by insects at Wadi Halfa and roasted by the early morning sun in Khartoum. The relatively cool, 75°, bearable sun and gentle breeze in Nairobi was most welcome. Fitness really counted here as we were 5,500 ft. above sea level.

There followed fifteen months of varied activity in magnificent country, full of potential, temporarily arrested by conflict and unrest. The battalion was primarily based on the foothills of Mount Kenya when I arrived, but moved to several parts of the country during my stay, which meant I was able to see at close quarters large areas of the Rift Valley, the Aberdare mountain region and the Highlands where many large farms had been established.

Comparisons were inevitable and I had plenty of time to observe the differences between my previous experience of farming in temperate regions and what took place here. The emphasis on hand labour, especially by African women. The vast fields of pyrethrum and groundnuts, coffee and cotton.

As a schoolboy living amongst hop fields, I had been as ready as any to make fun of London-born children coming to Kent to pick hops with their parents expressing disbelief when they saw cows being milked. 'In London we get our milk from bottles' they would remark. Now it was my turn to stare in almost disbelief at what my Geography master must have told me many times before—coffee comes from berries that grow on bushes—peanuts grow in the ground—cotton grows on shrubs. And the bananas! I offered the equivalent of 1p. for some bananas at a local market and was inundated with a hand of bananas which must have weighed five or six pounds!

There were only five R.E.M.E. personnel attached to the

battalion, one sergeant, three craftsmen and myself, by this time a corporal. We were responsible to the Rifle Brigade second Lieutenant who ran the motor transport section and was in charge of all vehicles and drivers. The vehicle compliment was made up of Land Rovers, three ton lorries of various types, an ambulance and a ponderous but reliable recovery truck.

My main task was to see that the vehicle inspection, maintenance and repair system worked smoothly. This involved a lot of documentation, vehicle inspections, supervising the craftsmen on repair, carrying out repairs myself and making sure that the drivers did their maintenance regularly.

Most of the drivers were only just eighteen, flown straight from England after a two week drivers course and given charge of a three ton lorry on arrival. Almost all roads were of earth construction, steeply cambered and liberally dotted with deep potholes. Vehicles and drivers frequently failed to cope with these conditions especially in the rainy season when the road surface became as slippery as the M1 on a frosty morning!

Ground adhesion, wheel slip and traction are all very important factors when tractors are used to pull implements and machinery. When I came to consider these problems in later life as an agricultural engineer I found that my experiences with vehicles in Kenya gave me an excellent appreciation of the problem.

Vehicles often simply became stuck in the mud on the side of the road when recovery was relatively simple. On occasions however, difficulties were encountered on steep escarpments or just as a river was about to be crossed. Then the final resting place of vehicle and driver was much more difficult. News of such accidents seemed to have the habit of arriving in the early hours of the morning but always the recovery team, either the sergeant or myself and two craftsmen would set off immediately the exact location became known. Recovery could take three or four days including a day each way travelling, so there was a considerable sense of adventure associated with each call.

Wild life becomes most active at dusk and dawn. It was not uncommon to round a bend on a lonely track having to bring the seven ton vehicle rapidly to a halt to allow a herd of twenty or so elephant to cross the track on their way to a nearby water hole. Rhinoceros and buffalo were also treated with the greatest respect by the recovery team.

I can recall going to recover one almost new Land Rover that had actually been attacked by a rhinoceros no doubt because of the impatience of the driver. According to him, the radiator grille and half the radiator was plucked out of the Land Rover as easily as a winkle from its shell and it took several seconds of head tossing before the rhinoceros could free the grille which had become stuck on its horn!

It was on these vehicle recovery missions that I learnt how important teamwork really is—not just on the sportsfield which had been the limit of my experience so far. On the farm it was there but I hadn't recognised it as such. The trips were frequently difficult and sometimes dangerous. If anything went wrong I would be held responsible in the pedantic but necessary Army tradition. At the time of recovery however, every man was responsible for his own job whether it be positioning anchor plates, driving in anchor pins, selection of the right blocks and pulleys, or operating the winch smoothly.

There was plenty of engineering involved as well. We had learnt about mechanical advantage, velocity ratio and resolution of forces in physics at school. It came up again at the technical college and here I was 6000 miles from home using the knowledge in a very practical way!

During my stay in Kenya I spent as much time as possible outside the camp. Weekend trips could be made to towns and some of the larger farms in which I had a particular interest. I even managed to get to the Kenyan equivalent of our own Royal Show held in Nairobi at which much of the farm equipment and machinery available locally was displayed and demonstrated. It was the machinery that I had not seen used in England which interested me most. Blades were exhibited for mounting behind tractors to produce terraces, a method of stepping the soil on slopes to avoid erosion during the rainy season. Groundnut lifters and hand tools for use by the Africans were quite new.

A four week safari by car into what was then known as Tanganyika enabled me to see the farming techniques in this part of the world as well, although it was the majestic flat topped snowcapped Kilimanjaro and the abundance of wild life in the Ngorongoro crater which impressed me most. I had lived in a tent on the foothills of Mount Kenya for six months. Although only a few miles from the Equator this mountain, over 17,000 ft. high is also snowcapped, but after that length of time was quite wrongly taken for granted. One sport which the R.E.M.E. group founded was to pit our engineering ability against the rarefied atmosphere by tuning Land Rover engines to see whose technical skills would carry us highest up the mountain tracks! A sport at the time but loss of engine power due to working at high altitudes is an important consideration in many parts of the world for farmers as well as tourists.

By this time I was beginning to realise just how much National Service was opening my eyes. Imagine my delight when I learned that with only four months of my two years to serve, I was to accompany the Rifle Brigade on their next assignment in Malaya. We sailed for Singapore via Colombo.

This additional and unexpected experience demonstrated to me the enormous differences which exist between countries. Johore, the State in which I spent most of my time is also close to the equator but what an effect altitude has on climate! The temperature was often only slightly higher than parts of the Kenya highlands but the much higher humidity made life very sticky. Mosquitoes seemed to thrive in larger numbers and scorpions provided considerable excitement and panic on occasions, especially amongst the local people.

My job was much the same as in Kenya although the range of vehicles was extended to include special purpose vehicles such as tracked and balloon tyred carriers. The problems of conveying Army equipment and men overland have much in common with agricultural requirements as I was to learn later. I again lived a very full life seeing and doing as much as possible in the time. It would have been relaxing to have spent my remaining seven days leave lazing on one of the beaches at the leave camp. Instead I accompanied the battalion advance party to Kuala Lumpur to see more of the country, driving through the rubber plantations and more mountainous regions of the north. Several of my colleagues thought I was mad! I considered the trip exhausting but well worth while.

In Kenya I had played sport a little but opportunities seemed to be limited. Not so in Malaya where a lot of cricket was being played, and I couldn't resist the temptation to join in. It was like playing and taking a Turkish Bath at the same time on occasions.

The journey home provided yet another unexpected experience thanks to the Suez crisis. Our boat was turned back at the entrance to the Canal and despite a brief stop at Mombasa, food supplies had run out by the time we reached Cape Town. It took a few days to rectify the situation during which time the South Africans were most hospitable. Quite by chance I met a farmer on the quay-side who took me to his home and showed me several aspects of farming methods in this part of South Africa.

Sheep farming was the most common activity. My first thoughts were that the agricultural engineer would have little to contribute here. During discussions however, it was pointed out that animals need food, water and in some cases shelter. Many sheep farmers tried to produce as much food as possible for the sheep on their own farms. This meant growing some crops with the aid of machines which would also be used to help manage the grazing land properly. Adequate supplies of water had to be maintained and it was sometimes necessary to pipe water to outlying parts of the farm. Water pumps powered by electric motors, engines or even by wind machines were used.

My two year conscription was thus completed. I was glad to be back in England, but regretted little in a phase of life which had broadened my outlook considerably. I had plenty of time to think about my future during the six-week journey home, and had reached the conclusion that further study towards preparation for the National Diploma course would be worthwhile. The only question which remained was—How?

4

PRACTICAL EXPERIENCE IN HORTICULTURE

My parents had moved to another part of the County while I was abroad. As we had agreed that I should try, if possible, to live at home during this year—the cheapest digs in the world!—I hunted around for a Technical College in the area which would accept me on to the final year of the Ordinary National Certificate (ONC) course.

It looked like being a messy year as I also had to find some sort of job to fit in with my studies. I had no savings, couldn't qualify for a grant and felt it quite unreasonable to live off my parents at the age of twenty-one.

At this stage I still wasn't absolutely certain about agricultural engineering as a career but my attitude towards study had grown much more serious since those final years at school. I had begun to see life as a series of opportunities to be accepted or rejected—if recognised. My ability to recognise opportunities had been slow to develop in the past, so in order to make up for lost time I set myself a really stiff programme for this year.

I went to the Technical College for two days and five nights each week. The Head of Engineering doubted whether I would keep it up but was prepared to let me try. I had decided to attend classes in the third year ONC in both Mechanical and Electrical Engineering.

For the first time I had the bit between my teeth, anxious to make some real progress towards becoming an engineer—of this much I was certain.

For the remaining three days, to make up a full working week,

I took a job as a lorry driver and general labourer on a market-garden enterprise run by an uncle not far from the College. There was plenty of hard work to be done but being out in the fresh air helped to make it acceptable and often enjoyable.

A dominant feature of market garden or horticultural work is the amount of hand labour involved. The crops are often costly to produce because of this. Income can vary widely from year to year on a particular crop if weather conditions are not favourable. Efforts are made to compensate for unsatisfactory weather conditions by irrigation and for some crops weather effects are minimal since an artificial environment is provided in the form of glasshouses. But someone has to provide the means of irrigating the land and erecting the cover for crops without restricting light which is vital for growth of plants. This has largely become the responsibility of agricultural engineers. Without fully realising it at the time I was gaining experience of yet another aspect of the agricultural engineer's work.

Tomatoes, cucumbers, lettuce and some early strawberries were all grown under glass. These had to be attended to regularly and eventually picked by hand. This work was ideal for getting a good suntan and a far cry from gathering sprouts in the fields by hand in heavy frost which was a regular chore in the winter months.

Once gathered the produce would be loaded on to lorries and delivered to shops in nearby towns for sale to the public. On occasions it would be necessary to take some of the produce to the London markets when supplies were plentiful.

Despite the travelling and ever increasing involvement in the running of the market garden I managed to cope with my three courses at the Technical College. I did not find the work difficult which was just as well, because there was precious little time for study outside the classroom! Electrical engineering was strange at first but a slowly improving knowledge and understanding of mathematics was a great help. Studying engineering without mathematics is like trying to overhaul an engine without any spanners. It is a universal tool which enables so many subjects to be tackled smoothly and efficiently.

The examination results were very encouraging to me. At last I seemed to be getting somewhere with my studies. The year which had started out so uncertainly was proving to be quite successful. There were times of course when I wondered whether it was worth the effort—especially when the cricket season got under way. Being committed to study five evenings each week when my friends were off enjoying themselves required dogged application and dedication that I would never have believed myself capable of!

Once the examination results were known I decided to confirm my intention to study next year for the National Diploma in Agricultural Engineering. There were still odd moments of doubt but these were quickly dispelled by a letter of final acceptance and Joining Instructions for the next session at the Essex Institute of Agriculture, followed by confirmation from my County Authority that a grant had been approved.

Almost three months remained before I was to present myself at the Institute. During this time I took digs close to an agricultural engineering firm selling and installing hop picking and continuous drying machinery in the county. As a fitter's mate I earned a useful sum of money working about sixty hours each week on many farms in the county. The acquired capital provided a worthwhile supplement to my grant the following year.

Hops are a specialised crop grown commercially only in Kent, parts of Sussex and Worcestershire. I was proud to be associated with increasing productivity of such a vital constituent in our national beverage—beer! Increasing labour costs of hand picking precipitated the development of huge machines 60 feet long and 20 feet high, located in farm buildings. The hop bines are brought to the machines by tractor and trailer for removal of the cones and separation from leaves and stalks.

Each machine replaced about 200 pickers which meant that thousands of Londoners and their children had to seek alternative occupations and holiday arrangements during the month of September when the hops are harvested. Progress in mechanisation never proceeds in the absence of side effects—a fact which is often of paramount importance in developing areas of the world—but more about that subject later on . . .

5
BACK ON COURSE:
THE NATIONAL DIPLOMA
IN AGRICULTURAL ENGINEERING

SO IT WAS TO BE Agricultural Engineering! My first formal introduction to the subject was about to begin. I arrived complete with trunk and suitcases and eventually found my Hall of Residence and very comfortable study room.

The Institute, set in flat country on heavy Essex clay, was primarily concerned with Agriculture with a secondary, but considerable Horticultural interest. The committment to Agricultural Engineering came third, just above other 'fringe activities' like Poultry Husbandry and Beekeeping.

I was surprised to find only fifteen students from a total of about two hundred on the Agricultural Engineering course. I fully expected three or four times that number especially in view of all I had seen in my travels that needed to be done by agricultural engineers. After spending a few years looking at the profession from the outside, I had misjudged its size and state of development. Clearly I had joined a branch of engineering still in its infancy. Would it grow or decline? Was I doing the right thing by becoming involved or should I have gone on with one of the more traditional branches—Mechanical Engineering for example?

Each new phase of one's career seems to be accompanied by doubts and fears. Military service started off by being a potential waste of two years but worked out very well subsequently. Perhaps there is some substance after all in the saying 'Life is what you make it'. A profession is not created overnight, it evolves and progress depends very much on the people in it. The idea of being in on the ground floor appealed to me and I soon found myself getting down to the programme of lectures, seminars, visits and practical classes with enthusiasm.

Entry on to the course was either via 'A' levels, the National Diploma in Agriculture or, as in my case, the Ordinary National Certificate in Mechanical Engineering. All entrants were required to spend at least one year on practical training before taking the course but this was sometimes restricted to only one or two aspects of farming or engineering.

My varied practical experience of both farming and engineering was frequently used as an aid to understanding and generating personal interest in many parts of the course. I had seen crops suffering from lack of nitrogen in the soil, damaged by pests and stricken by disease. Animal housing and feeding techniques had also been observed on the farm. Ploughs mounted on wheeled tractors and trailed behind crawler tractors had been used, adjusted and repaired.

When the provision of plant nutrients was considered, the control of pests and diseases discussed, the points for good design of animal shelters enumerated and soil forces acting on plough bodies analysed I was able to refer back in my mind to these experiences. In this way I am sure that I gained more from the course than those who were approaching the material for the first time.

Gaps were filled in and knowledge extended. With a broad base upon which to build it all seemed so much more significant and worthwhile.

Being a small group we quickly got to know each other and were soon swopping stories about careers and experiences of life. We also had plenty of opportunity to meet students studying Agriculture and other courses offered at the Institute. My closest friends were Tim and Tony, two course members who, like myself, had decided to take Field Engineering as a special subject.

Tim was a tall, long haired ex-public schoolboy, with a charming manner and polite nature, but frequently seemed vague and remote. He was never seen hard at work but always had a thorough grasp of the coursework. His briefcase was virtually taken over by piano scores with just an odd corner for lecture notes. Every spare moment seemed to be spent playing concertos and sonatas on the grand piano in the main hall rather than looking up references in the library. Tim was a good example of the small number of individuals who can follow a course and pass examinations with the minimum of effort and study. This approach would have been disastrous for me!

A meandering path had been followed by Tony which eventually led him to a study of agricultural engineering. The preceding years had also included travel overseas during National Service. However, the most important factor which had drawn the three of us together was a mutual interest in the arts. We were each agreed that specialisation was necessary in order to get a job, follow a career and progress. It was also strongly desirable if not essential to maintain and develop broader interests for a fuller life.

Several enjoyable evenings were spent together listening to concerts at the Festival and Albert Halls. We made many visits to the Art Galleries in London and spent hours and hours discussing our interests and fears of becoming lost without trace in the industrial jungle. I often made the point that to be aware of a problem is half way to solving it. Only those who are not aware of the dangers of becoming 'lost' or deeply entrenched in the metaphorical rut are likely to suffer the fate which we had become so concerned about.

The choice of Field Engineering as a special subject had been pretty arbitrary. I was interested in the design of farm machinery but what attracted me about Field Engineering was the possibility of working overseas. I had by now acquired a taste for travel and would have liked to go back to Africa without wearing a uniform. It seemed to me that the description of the subject as being 'A study of the control of soil and water, including that applicable to the agriculture of dry regions overseas' was well worth looking into in more detail.

I had seen dry regions in northern Kenya where water was valued highly, especially by those trying to grow crops. Many relied solely upon natural rainfall which, over the year, was probably the same amount as in South-East England but it all fell in a few weeks followed by several months of blue sky and baking hot sun. The soil which was not washed away would be baked hard. It would shrink and then crack. These cracks would widen to gulleys during the next rainfall and some would eventually end up as crevices several feet deep. Where the soil remained loose, drying winds would blow the top layers of soil away exposing inert and barren earth, on which nothing would grow. Plant roots help to bind the soil together and if absent the process of soil erosion occurs much more readily.

If the land is sloping, then water flows over it more rapidly—and always in the same direction thus speeding up the erosion process. Motorway bankings are quickly planted with bushes or trees and sown with grass for this very reason. The slope is often terraced when crops are to be grown.

Once again I was able to relate what I had seen in practice to the methods and techniques given in lectures. How to conserve soil and water; when to consider irrigating from above through pipes or at ground level in channels.

In some cases of course there may be too much water in the soil which has a harmful and, in the extreme, fatal effect on most plants. This occurs especially where soil is heavy and retains water readily and where the level of water beneath the soil surface, the water table as it is called, rises frequently due to flooding.

The design and construction of the ever widening range of equipment available to the agricultural engineer for helping to solve these vitally important problems was also considered in detail.

I was surprised to learn that problems of erosion and water shortage were not always confined to hot and dry parts of the world. There are for instance, serious examples of soil erosion in the Fens and most of the Eastern half of England can and does benefit from irrigation during the summer months.

The Field Engineering group had some fun designing and building a bridge over a stream several feet wide running through the College grounds. Individual designs were initially produced from which a final arrangement was agreed in discussion. It was to be a single span bridge constructed from timber resting on concrete pillars. The loads which it would have to carry were estimated from a description of the expected use. The idea was to introduce us to problem synthesis where, unlike problem analysis, gaps exist in the information necessary to obtain a solution.

In practice, synthesis often has to be applied to problem solving which means that experience becomes a vital factor. The range of experience present in the group combined with occasional failure to apply the theory correctly led to a range of designs suitable at one extreme for supporting little more than the weight of bridge construction material itself, to structures over which tanks could have been safely driven!

The final design lay somewhere in between. Materials were obtained and after several soakings in the stream, the group finally completed the exercise—assisted here and there by Institute technicians. Those who originally designed a much more robust structure walked across with caution whilst the light structure men proceeded with gay abandon in the full knowledge that it was really overdesigned anyway. So far as I know the bridge is still standing to this day. Perhaps it may eventually become preserved as being of historic interest!

Several other topics were considered important to the training of Field Engineers including building of earth roads, working with concrete, provision of farm services and surveying. I always enjoyed surveying since to me it was a neat and precise method of making maps, although there are of course several other applications of this technique to the work of the Field Engineer. In the workshop I had used micrometers capable of measuring lengths and diameters in the order of ten centimetres to the nearest quarter of a millimetre or less. Surveying instruments were at the other end of the measuring scale dealing with a hundred metres or more at an accuracy of around 10 millimetres.

The practical classes were always interesting and I shall never forget our narrow escape when we attempted to assess various methods of removing tree stumps as part of land clearing techniques. Winching them out was relatively straightforward. In our enthusiasm for making an equally good job of the blasting method I am sure we used enough explosive to demolish the Post Office Tower! The roots attached to the wretched stump offered negligible resistance to our charge and it was launched Apollo fashion into a neighbouring field! Fortunately, our close adherance to the safety precautions ensured that I and my colleagues survived to tell the tale, suffering no more harm than temporary loss of hearing and no doubt a poor assessment for the exercise.

Lectures and practical work were usefully supplemented by visits to local industry, mechanised farms, distributers of machinery and the like. I attended my first Smithfield Show which, together with the Royal Show, constitutes the Mecca of all agricultural engineers intending and practising. Tractors and a vast range of machinery stand gleaming and unfamiliar owing to the marked absence of mud and unusual presence of chrome and bright paintwork.

With so much going on time passed quickly. My priorities had changed. Had I not been elected captain of the College cricket side, I might have actually missed a few matches! As it was, practice was strictly limited to one evening a week and no practice at all during the three or four weeks preceding exams. My devotion to duty was rewarded. Another examination

AGRICULTURAL ENGINEERS OF THE FUTURE

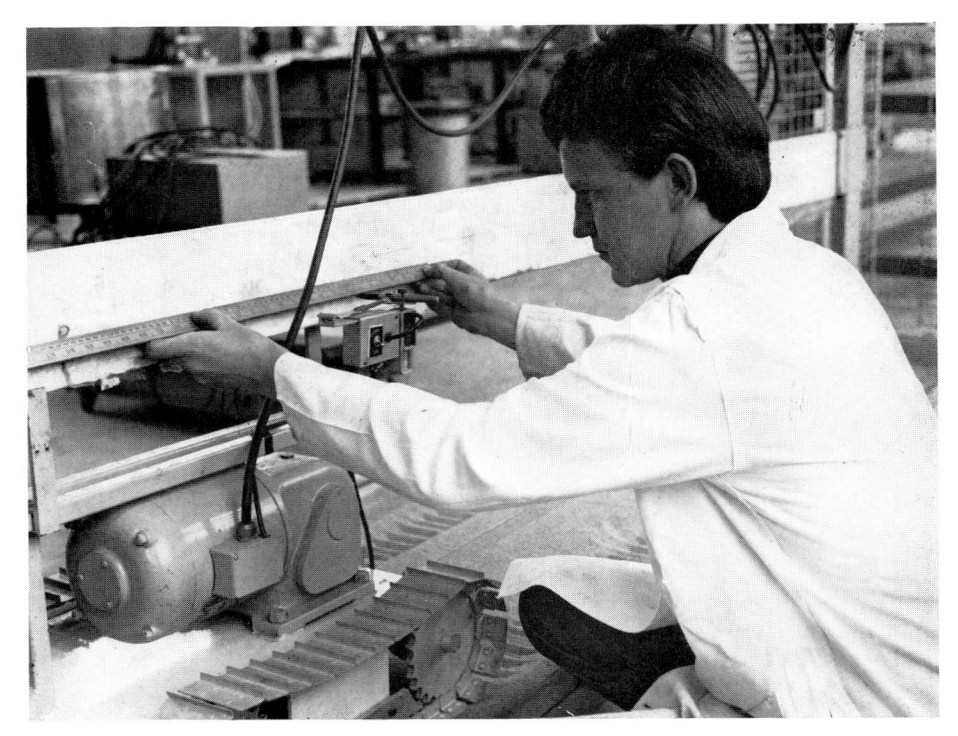

1 Experiment with a model tractor at the National College, Silsoe

2 Students assessing the performance of a mechanical potato planter in the lab

3   Design: a blackcurrant harvester which has now gone into production

AGRICULTURAL ENGINEERING: A BROAD SUBJECT

4   Control of the Environment: a mist-propagator and overhead heating tubes stimulate growth in a seedling nursery

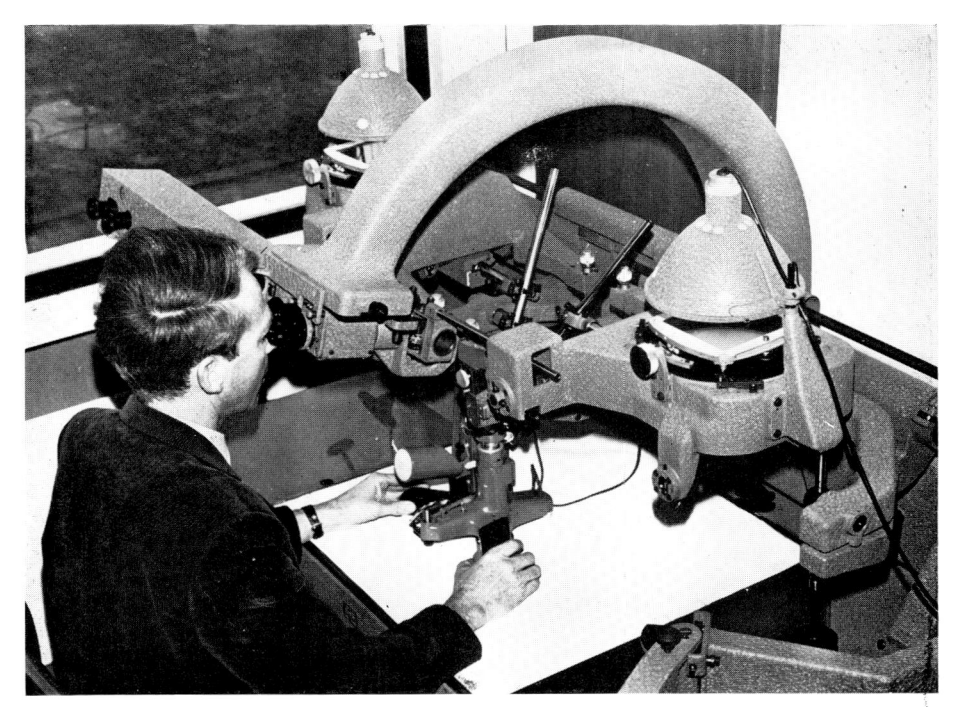

5   Field Engineer: stereoscopic plotter used to prepare contour maps from aerial photographs for planning irrigation and drainage schemes

6   Research: study of noise levels in a tractor-operator's cab

MANUFACTURE OF FARM
MACHINERY

7   At Home ... combine harvester assembly line

8   Final stage in farm tractor assembly

9   And Overseas . . . a simple mechanical cultivator takes shape in Tanzania

10   A bicycle-operated separator for cereal crops

MECHANIZATION

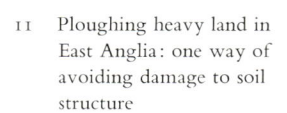

11 Ploughing heavy land in East Anglia: one way of avoiding damage to soil structure

12 Cultivating rice paddies in Ceylon

13 The start of the day's work on a farm in Japan

14 Harvesting sugar cane in Jamaica

15 An Iraqi irrigation project partly designed by Agricultural Engineers to lift water from the Tigris to benefit agricultural land

IRRIGATION AND LAND DRAINAGE

16 Laying tile drains in a field to lower the water table

success was obtained and my unbeaten record in battles with examiners since leaving school was preserved.

As the examination period approaches thoughts begin to turn seriously to what happens when the course is over. For most of my colleagues this meant either a straight job with a manufacturer or distributer of farm machinery or an appointment for two or three years overseas normally in a developing country.

Tim, Tony and I, aware of the risks of becoming perpetual students, each felt a lack of fulfilment and a wish to take our studies further before seeking an appointment. Although the National Diploma course had been interesting and of value, it was general in nature and lacked depth, particularly in the analytical aspects of the engineering subjects. For many jobs this additional depth would not be essential. We did not believe however, that our newly acquired qualification would enable us to compete on equal terms with contemporaries trained at a university at degree level. We had, therefore, reached the end of the academic road in agricultural engineering since the only degree level course at the time was the M.Sc. at Newcastle. A Bachelor's degree was necessary as an entry qualification for this course. The National College and its Bachelors' degrees in Agricultural Engineering were still five years away and although we felt the need for such a course very strongly, at the time it just wasn't there.

In the event, Tim decided to study for a first and higher degree in agriculture as he already had a National Diploma in the subject. Tony went off to get some belated A levels and then to University for a degree in Civil Engineering. I eventually studied for a Diploma in Technology in Mechanical Engineering later to be converted to a degree in that subject. I have previously mentioned opportunities and I now add one more point to my philosophy—if the opportunity that you seek doesn't exist, then create one!

Since the next stage in our respective careers would not be

commencing for two or three months we decided to spend this time together by taking to the road in France and Belgium, ending up at the World Fair in Brussels. It was a terrific feeling to be free and without a care in the world! I would certainly not choose it as a way of life, but as a means of meeting ordinary people as they really are, it was second to none.

We introduced ourselves to several French farmers and farm workers gathering the harvest who never failed to be amused by our strange dress, haversacks with sleeping bags rolled up on top, and an outmoded use of their language! A sense of humour, genuine interest in people and friendly approach crosses pretty well all barriers. We were often invited into homes for wine, bread and cheese, or perhaps soup. Sometimes we even slept with a roof over our heads. Barns, school rooms and a camping site broom cupboard were all sampled.

Industrialisation has made heavy demands upon available land in Belgium. Agriculture has therefore diminished to an extent which makes the country heavily dependent upon imported food.

There was a very different situation in France. Much of the country was rural and many farms were small units often run by one family. Mechanisation becomes an expensive business in these circumstances and was therefore limited. Another reason for lack of mechanisation in the wine areas was the special problems presented by the crop. Agricultural engineers must always take full account of the agricultural and processing requirements when developing machinery. Specially designed narrow tractors and implements are available which can cultivate between the rows of vines. The problem of mechanically harvesting the delicate grapes has not yet been solved.

The World Fair, which was our primary objective, proved to be well worth the journey from Paris sitting in the back of a lorry on top of several tons of loose grain! There was so much to see including a wide range of farm machinery from the Soviet

Union. To three young men who considered themselves fully fledged agricultural engineers this exhibit was viewed through the eyes of professionals.

The English speaking exhibitors, confused by our unusual apparel and somewhat unkempt appearance, addressed us with caution. Our courage weakened only once when a tall, heavily built man, wearing a long raincoat and a broad-rimmed trilby hat pulled well down over his eyes, commanded us to come and inspect the Russian designed 'hanging machines'. To our great relief, this was his way of describing the hydraulic lifting device commonly fitted to the rear end of tractors to carry and control mounted equipment!

The tractors were huge and much more heavily constructed than those I had previously come across. Many were fitted with substantial cabs to protect the operator from adverse climatic conditions—a feature not often seen in England at that time. The comfortable seat and accessibly mounted radio and vacuum flask were also impressive. It would seem that the Revolution had had far-reaching effects.

Perhaps my new appointment which was to be with Massey-Ferguson would give me the chance to help provide some of these features for British farmers.

6

DESIGN TRAINEE WITH MASSEY-FERGUSON: DEGREE STUDIES IN MECHANICAL ENGINEERING

THE DECISION TO PURSUE a more advanced study of mechanical engineering was taken because I wanted to be better equipped to try my hand at the design of farm machinery. I could probably have made some attempt at design using my previous experience at College and on the farm. There did, however, seem to be gaps in my technical training. The National Certificate courses had given me a useful introduction to what was involved, but that was all.

At this time I was hoping that my departure from the Field Engineering aspects of the subject would be only temporary. I had found this work absorbing and the time spent out in the open was particularly appealing. If I wanted to find out more about this work then I should have to spend some time studying Civil Engineering. For my current interest in design it seemed that Mechanical Engineering would be best.

From a look at several University Prospectuses in the Library I soon discovered that 'A' levels were almost essential for entry on to first degree courses. What I was looking for, was a degree-level course in Mechanical Engineering for which my unusual bunch of existing qualifications would be acceptable.

After a good deal of searching I came across the Diploma in Technology courses run by Colleges of Advanced Technology. These courses were recognised as being of honours degree standard and had two methods of entry to allow for non 'A' level people like myself. These Colleges have now become Universities and the courses degrees, but I wasn't really interested in the magic words and letters at the time—just the right course! It is still possible to get on to these courses (and also several other degree courses) with good O.N.C. results.

One important feature of the Dip. Tech, as it became known, was the alternating periods between College and Industry—a sandwich course in fact. The length of the periods varied from three months to a year. Having spent much of my previous training on a sandwich basis, the idea appealed to me. People can get out of touch with the outside world if they spend too long in the rarefied College atmosphere. This applies equally well to both students and staff. Time spent in College or University should be a means to an end—not an end in itself!

Government or Local Authority grants were not available to me for this course since I had already received one grant for the N.D.Agr.E. The usual approach for O.N.C. entrants was to be sponsored by one's employer, so my next step was to find an employer who would be prepared to foot the bill for my further education.

Not surprisingly, it was difficult to find a Company willing to do this. My previous career record was neither standard, nor impressive. Personnel Officers and Managers in Industry were more inclined to recruit trainees from established sources, primarily school leavers with good 'A' level passes.

My method of tackling this problem was to prepare a case which justified my past record—to my satisfaction at least. My previous interviews had taught me that employers not only look for a satisfactory academic record, but also include in their assessment personal qualities such as character, personality, motivation and interests. Perhaps I might convince an employer that I was worth a risk on the basis of these personal qualities.

I wrote a long and detailed letter to twenty firms setting out my case as clearly, carefully—and honestly—as I was able. I included reasons for wanting to continue my studies and a paragraph about my future hopes and ambitions.

The letter was handwritten, neatly and clearly set out, and an individual copy prepared for each firm. It seemed a laborious and hardly worthwhile exercise at the time. I had, however, been advised that many Personnel Officers are influenced by the form and content of applications for appointment—more than one would imagine. As a member of several interviewing committees in recent years I have certainly found this to be true. The letter—often preferred to a form—is the first contact with the applicant and can leave lasting impressions—good or bad.

The response was varied and often disappointing. I understand that there is a northern expression 'When in doubt, say nowt.' This was quite appropriate to this exercise, as several did not reply at all. Others politely, but firmly declined the opportunity of paying for my proposed course. There were those who replied that they had never heard of the N.D.Agr.E.—which had a withering effect on my confidence. I was naïve enough to think that simply everyone should have heard about my recently acquired qualification!

There were a few interviews, but none really raised my hopes of getting on to the Dip. Tech. course until I was invited to visit the Massey-Ferguson offices in London. The Personnel Officer was interviewing a number of graduates for the graduate training course and somehow I had slipped into the net. Although he didn't consider me suitable for his present requirements he did promise to discuss my one-sided proposition with the Chief Engineer in Coventry. About ten days later I was delighted to receive a request to attend for interview at Coventry and somehow felt that the visit was going to be successful.

I was not to be disappointed this time. The Chief Engineer listened carefully to my story. I sensed that at least some of the points I was making were accepted and his two senior colleagues present also appeared to be 'on my side'. An offer was made towards the end of the interview. The Chief Engineer, after summing up what I had had to say, pointed out that my industrial

DEGREE STUDIES IN MECHANICAL ENGINEERING 63

experience was still limited. He would be prepared to offer me a job for one year with his Company as a Trainee Design Draughtsman. I would work in the design office learning to produce and interpret drawings, and also to gain some experience of the other activities in the office. During this time I would be allowed one day each week to continue my National Certificate studies at the local technical college. If at the end of this period I liked the look of the Company—and the Company liked the look of me—then I would be able to continue on to the Dip. Tech. course at Birmingham.

This was excellent news! I thanked the interviewing panel for this tremendous opportunity to get into industry and left the meeting walking on air. There was the slight problem of an additional year which I hadn't bargained for, but apart from this the offer was perfect. Subsequently I found this experience on the drawing board invaluable, so the offer in fact turned out better than I was able to judge at the time.

The Assistant Personnel Officer arranged for me to stay temporarily in his own digs. This proved a very useful arrangement for the first few weeks as I quickly got to know about the Company organisation and staff during our many conversations over dinner and in the lounge of the well-appointed detached house on the outskirts of Coventry.

I had spent many hours at home in Kent wondering what to expect when I began this new phase of my career. Would it work out successfully? Did the Company really mean to give me what seemed to be an opportunity almost too good to be true?

My first day in the design office was spent meeting people and visiting the various sections working on the design and development of different agricultural machines, including tractors. The Chief Engineer and his senior staff were just as friendly and helpful as they had been at the interview. On reflection, this is not particularly surprising but at the time I was still on my guard wondering whether their attitude might change once I became an employee. My fears proved groundless.

Each section consisted of a project leader, one or two design engineers, a senior draughtsman and usually three or four assistants. I was to join a section dealing with special projects as a junior assistant.

The Company was undergoing a major change of identity —a feature which often occurs in industry today. Just prior to my arrival, the Company was still run by the late Harry Ferguson, an outstanding agricultural engineer. Ferguson's major contribution to farm mechanisation was made in the 1930s when he designed a small light tractor which had a performance equal to its heavier and more powerful competitors.

This was made possible by a three-point attachment of implements at the back of the tractor (see diagram). The two lower links connecting tractor and implement were controlled by a hydraulic lift mechanism and instead of simply being pulled over the ground, the implement was suspended by the links. This meant that part of the weight of the implement was carried by the tractor. In this way the force between the driving wheels and the ground was increased which enabled the tractor to get a better grip and produce more pull or draught as it is called.

The upper or top link was connected to the tractor by a strong spring. Any change in length of this spring caused the hydraulic pump control valve to move. If the implement went too deep the force in the top link increased. The spring would therefore be compressed and, through a linkage, move the valve so that more oil would be pumped into a cylinder which lifted the other two links. This raised the implement and allowed the spring to extend until the valve closed, stopping the implement from being raised further.

If the implement wasn't deep enough then the spring would extend a greater amount opening the valve, but this time to allow oil to escape from the cylinder. The links and thus the

DEGREE STUDIES IN MECHANICAL ENGINEERING 65

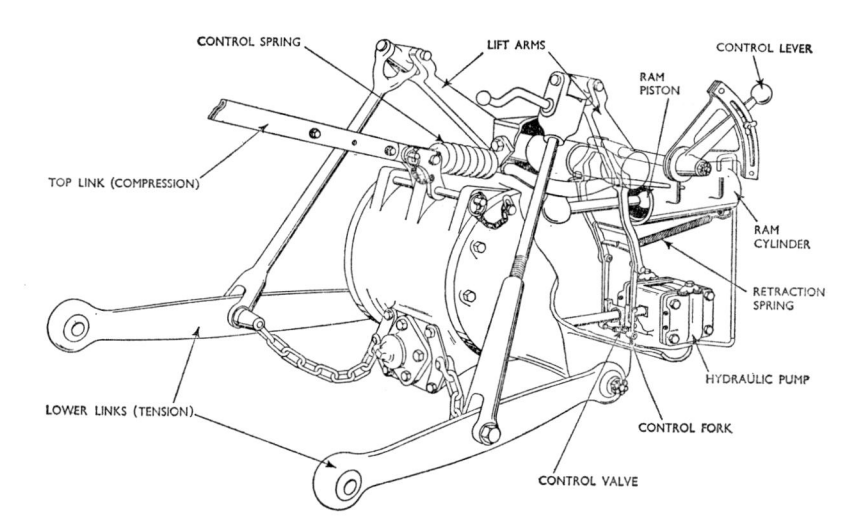

Diagram 1. Basic three-point linkage for attaching mounted implements to a farm tractor

implement would be lowered. The 'neutral point' for the spring had to be fixed by a lever set by the operator. This is called a draught control system and is to this day fitted in a similar form to most makes of tractor.

The draught control system formed part of what was known as the Ferguson System. The idea was to provide the farmer with all the machines and power necessary to disturb the soil (ploughs) break up the large lumps of clods (cultivators and harrows) to sow certain seeds (potato planter) to thin plants to give just enough room for development (thinners and gappers) and to keep weeds down (light cultivators and hoes). Each implement was designed to be the right size for the tractor.

One unusual feature of Harry Ferguson's company was that it only designed and developed the machines—it didn't make them. Thousands of tractors were produced by other Companies working under a licence—not without problems! Many implements were also made by the same method.

Massey-Harris, a large firm based in Canada was at that time making combine harvesters in Scotland for use mainly in the U.K. The outcome of complex business negotiations was that Massey-Harris took over the Harry Ferguson Company to form a new organisation which eventually became known as Massey-Ferguson. In addition, the factory in which the tractors were made by the Standard Motor Company (now part of British Leyland) was also taken over. Other interests were subsequently acquired in the U.K. including the Perkins engine company at Peterborough. During my five years with the company the staff numbers grew from about 400 to over 10,000!

The brief description of the Ferguson System and company history serves to illustrate the background to my induction into industry. I soon realised that I was working with a rapidly expanding dynamic and progressive company—factors which contributed greatly to the sort of experience I was to acquire.

The Engineering Drawing classes that I had attended several

DEGREE STUDIES IN MECHANICAL ENGINEERING 67

years earlier gave me a good start on the drawing board which was allocated to me. I was allowed to settle in gradually by producing simple detailed drawings of small components on translucent paper for subsequent printing. This part of my training was not particularly stimulating but John, who used the drawing board next to mine, patiently and ably demonstrated the need to acquire the basic skills of draughtmanship before proceeding to more complex work. Considerable variation in line thickness and density on the paper didn't seem very important to me—until we came to printing! The correct use of different grades of pencil, giving the lead a chisel point to help maintain even thickness, clear and consistent printing and the use of correct symbols were just a few of the aspects which I had to master. It was several weeks before I realised that all my drawing board work at College had been done on cartridge paper which is rarely, if ever, used in industry. Perhaps a lot of the initial adaptation would have been unnecessary had the appropriate paper been used and some prints taken.

Despite the mundane nature of my work on the drawing board, the results were being used by the Company to help build up complete sets of drawings for new machines and to keep a record of modifications on existing machines. I was surprised to discover just how much modification and relatively minor development work went on in the design office compared with the design of new machines. In fact about three quarters of the design work was concerned with existing machines. This is not to say that modification exercises were uninteresting. The constraints imposed were often challenging and the specific reason for the modification created a sense of personal involvement. A power-take-off-shaft guard is clearly necessary to protect the operator from rotating parts and many are needed when over three hundred tractors are being made each day. A modified plough design to enable it to resist the abrasive soils in the tropics has a direct and easily appreciated purpose.

As the weeks passed, I virtually lost the sense of being the inexperienced newcomer and felt that I was accepted once again as a member of a team. I am sure that good fortune plays its part: the second chance given me by the Personnel Officer in London, the chance to be interviewed by the Chief Engineer and his decision to offer me a one year probationary period, being accepted by John as a young man worth encouraging. My response at each stage of this sequence of events was no doubt also of great importance, but without opportunity, response is of little use.

I never discovered whether John was formally selected to 'take me under his wing' for the year. Perhaps he was just asked to set me off with the basic principles of draughtsmanship and the rest developed informally. In the event our association and friendship quickly strengthened and has continued for many years. I certainly believe that much of the experience gained during my probationary year was directly as a result of his efforts. My willingness to listen and learn probably helped but the presence of a patient, experienced and helpful adviser made all the difference.

Proceeding to the first year of the Higher National Certificate Course enabled me to maintain an interest in academic studies. I regarded this course as an insurance policy. Should I not be accepted for the Dip. Tech. then at least I would be in a position to negotiate with the Company for day release on to the final year of the Higher National Certificate. This was assuming, of course, that my probationary year would be extended!

As far as I was concerned I very much liked the look of the Company. There seemed to be plenty of career opportunities and the working atmosphere was certainly very friendly. Towards the end of the year the Chief Engineer called me into his office to give me the glad news that the Company would be prepared to sponsor me for the Dip. Tech. My objective had been achieved.

It only remained for formal application to be made to the College in Birmingham. At the College interview it was agreed that my previous studies and experience were sufficient to permit entry direct to the second year of the course. This concession was welcome as I was becoming increasingly aware of the time that my various courses were taking to complete.

Massey-Ferguson had for some time been training craft apprentices and graduates. The idea of accepting responsibility for the training of a young man actually to become a graduate was however, as new to the Company as it was to me! This fact coupled with the relatively recent introduction of the Dip. Tech. courses particularly appealed to me. The pioneering element was very evident. Its presence meant that there was plenty of challenge and room for manoeuvre. I can understand that to some this aspect would have been disturbing, but to me it provided freedom and the opportunity to mould the training period to suit my own requirements.

During the three month sessions at College I quickly came to realise that there was a marked difference in standard between my National Certificate courses and the Dip. Tech. I found that it is possible to study a subject—like mathematics for example—at one level and to pass the examination believing that there was little else to learn about the subject. Repeating the subject at a different level however, proved to be a very sobering experience! The syllabus might look the same but the treatment made the subject almost unrecognisable.

I did not become despondent about the situation. It seemed to me that a course of study is provided at a particular level appropriate to the sort of job you are preparing to tackle when qualified. Mathematics is required for many jobs from laying bricks to building spacecraft. The level and type of mathematics will certainly not be the same.

Despite the challenging nature of the College work I managed to keep my head above water and satisfy the examiners at the required times. I had started this part of my education believing that the academic work was the only part of real value. The training programme sandwiched between the College periods has proved to be of equal value to my subsequent career and therefore deserves a chapter on its own.

7

WIDER INDUSTRIAL EXPERIENCE: PRODUCTION AND MARKETING TRAINING

DURING MY YEAR in the drawing office I became increasingly aware that there was more to design than simply thinking up an idea and sketching it out on a piece of paper. The additional requirements extended far beyond the production of accurate drawings from which the parts could be made. I was continually finding out more about the calculations and analysis necessary from my lectures at Birmingham. Short periods spent with many departments at my company and other organisations gave me an insight into the enormous amount of teamwork that is needed to make a product available to the customer.

There is little point in designing a machine, or part of a machine, unless it can be made. This has to be done as easily as possible, and at the lowest cost. Nevil Shute once described an engineer as a person who can make for five shillings what any fool can make for a pound! There is much to be said for this description especially during times when high and rising costs are evident.

To be able to decide whether a particular design can be readily made, it is necessary to learn something about the machines and equipment available for this purpose. When preparing my programme it therefore seemed right to spend my first three month industrial period in the experimental workshops.

There are three main methods of using metal to obtain the required shape and size as stated on the drawing. One is to start with a standard shape slightly larger than required and then proceed to remove the surplus using special tools and machines, e.g. a lathe, shaper, miller, borer or grinding machine. The second approach is to heat the metal until it becomes plastic. It can then be shaped between hard metal blocks called dies using hydraulic or pneumatic pressure or repeated blows from a hammer (presswork and forging). The third method is to heat the metal until it melts and then pour the liquid into an appropriately shaped container. There are several ways of achieving this but the technique is generally known as casting.

I wanted to find out about each of these methods during my training. The experimental workshops were entirely concerned with metal removal techniques. I was given the job of producing several simple components on each of the machines available. This was done under the supervision and guidance of the skilled machinists whose job was to make prototype components and sometimes complete machines from the drawings supplied by the design office.

Before working with these men I had hardly given any thought to the important part they played in getting a designer's idea on to the production line. In addition to making the parts from drawings they were often able to assist the designer by discussing difficulties they had experienced in understanding the drawing, obtaining a particular shape or working with a tough type of metal probably selected for its strength. If problems of this type were not sorted out at this early stage serious difficulties could arise.

The Workshop Manager, who had worked with Harry Ferguson in Ireland, explained to me one day how some of his men approached the situation in an understandable, but inappropriate fashion. It is not easy for a man who takes a pride in his work and knows he can do a good job to admit that he is having difficulties. Rather than report an ambiguity on the drawing, some prefer to deduce the correct approach or assume some obvious but missing information. In this way they believe that they are making good use of their skill and experience.

What can happen in this situation is that an unsuitable or

WIDER INDUSTRIAL EXPERIENCE 73

incorrect drawing finds its way into the factory where the men and machines are making hundreds of parts and not just one. This is neither the time nor the place to check the drawing and a mistake could pass unnoticed until a whole batch comes to be assembled together with other parts and, if this happens, someone is really for the high jump!

While I was 'doing the rounds' of the machines in the workshop it soon became obvious to me why John had gone to great lengths to explain tolerances to me in the drawing office. It is not sufficient to simply state on the drawing that you require a hole for example, to be 20 mm diameter. The man whose job it is to make that hole wants to know how near he has got to be both above and below the stated size or dimension.

In real life it is not possible to make something to an exact size. The scrap which used to accumulate beside my machine verified this fact beyond doubt! Most holes are made with a drill. These come in standard sizes and if the hole required does not correspond to one of these, problems again arise. The moral here was to state wherever possible, standard hole sizes on the drawing together with as large a tolerance as could be accepted. After all, many holes are made to push bolts through and manufacturers of these take account of drill sizes anyway.

It was very encouraging to me when parts of my training fitted together in this way. The need to obtain a wide range of experience first hand was admirably demonstrated. It is possible to find reference to the importance of tolerances in textbooks, but for a thorough understanding and full appreciation there is no real substitute for experience—however brief.

During a one month visit to the Kilmarnock factory, where several types of machine were being made including combine harvesters and balers, I was able to see how metal was shaped according to the second method. Very large presses were necessary for shaping the brackets, covers and guards used on the harvesting machines. I had only seen these previously in their finished form either working on the farm or gleaming proudly on manufacturers stands at agricultural shows. In the factory they were stripped of all such splendour save for a few at the end of the assembly line and in the dispatch yard.

Walking down the assembly line and through the machine shop it was possible to identify the products in various stages of undress. I marvelled at the thought of the enormous amount of organisation that would be necessary to see that all went well in this complex line of activity. While it was not necessary for a design engineer to know how to perform all the functions of the people involved, it seemed that problems were referred back to the designer at so many stages during the process that I should at least be spending some time with as many of the specialists as I could during my training. I made a note of this for my future programme.

At the end of each three month training period I was required to write a report for my tutor at Birmingham. In addition, the Company Training Officer also had to send in a brief report. On each occasion he called me into his office to give me an outline of what he was going to say. His report was usually quite complimentary which always gave my morale a lift. Was this why he did it?

His favourite phase was 'Mr. May has demonstrated an ability to get on well with members of the department (or factory) that he has visited'. I found this ability, such as it was, extremely valuable. My temporary supervisors, guides and advisers were very varied in the way they received me and reacted to our normally brief encounter.

On reflection, I was really little more than an additional chore. So long as I recognised this fact and behaved accordingly, things went very well. By showing a genuine interest in the job and a willingness to work and listen to advice, I am sure that I secured maximum benefit from my visits. I had to make very sure of my facts and relationship with the individual before

WIDER INDUSTRIAL EXPERIENCE

pulling knowledge and information out of my College experience. I was the young and inexperienced College boy, and they were the experienced and skilled specialists. Like two reactionary chemical substances, they can be brought together, but only with the greatest care!

Strangely enough, my severest test in human relations did not occur in a factory, workshop or office. It came on the occasion of the Scotland-Wales rugger international at Murrayfield which I was invited to attend by the Kilmarnock Works Manager, an enthusiastic Scots supporter. The third member of our party was the Personnel Officer for the Scottish Factory who was equally enthusiastic about the sport but very Welsh and proud of it. As a mere Englishman interested only in the game I was required to display impartiality at least equal to that of the referee! I found my part in the post-mortem discussion during the return journey particularly demanding.

My next step was to obtain some experience of foundry work. Since the Perkins engine company had yet to join the Massey-Ferguson group, it was necessary for me to visit another firm for this part of my training. This fact in itself was valuable as I was able to see something of the organisation and procedures outside Massey-Ferguson. Many parts of the West Midlands are heavily industrialised, so it was not difficult for suitable arrangements to be made with one of the firms which supply M-F with castings.

What little I had learnt about this sort of work from lectures led me to expect a fume, dirt and dust laden building in which tough, cloth capped foundrymen sweated as they manhandled castings and equipment in the gloomy atmosphere. In fact modern devices such as fume and dust extractors, overhead gantry cranes and hand operated electric hoists made the work and conditions much more acceptable than I anticipated. Recently introduced and much cleaner processes were being used where possible, requiring skilled operation by technicians often dressed in white coats!

During my visit I was informed that when a casting is being considered by the designer he will produce drawings and send them to the foundry engineers for examination. Discussions then take place on design details, cost and ease of casting to arrive at the final drawings. It became clear that the designer needs to be able to talk the language and understand the problems of the foundry technician if they are to obtain a satisfactory component between them.

To supplement my machine shop experience I made a second visit to a company outside the M-F organisation. As we have seen, the basis of machining is to start with a standard shape of material, usually metal and often steel. Probably the most important open cast steel works in the country is run by Stewarts and Lloyds at Corby in Northamptonshire. During my few weeks visit I was able to see how some of the larger standard sections are produced using iron ore obtained from a few feet below the ground surface as a starting point.

As with the foundry visit, I could have been training to become one of several types of engineer to include such an item in my programme. This general training was, however, considered to be relevant and important to the training of an Agricultural Design Engineer. I have never regretted the experience for one moment.

During the following summer I began my series of short visits to the many groups of people who were helping to make tractors at the rate of over three hundred a day. The cost of running such an operation runs into several million pounds. If a mistake is made by the designer this could be repeated thousands of times in the course of a week or two. The company was planning to introduce a new range of tractors shortly and this project would be a good one to join when I became qualified. My chances would be considerably improved if I had a thorough understanding of the way in which the existing tractors were made.

Returning to the design office from time to time for discussion with the friends I had made there helped me to link the various parts of my training together. I was beginning to learn just how much the design of farm machinery involved. Effective teamwork was clearly the most important factor. The successful designers in the office were those who produced the sound ideas which could be readily put into practice by other members of the team. This was done by arranging the design details so that they made the best use possible of the available machines and equipment in the factory.

The designer is responsible for his design at all times and must be prepared to take account of any problems which arise in the factory or later when the design is in service and is being used by the farmer. This process of modifying the design as a result of new information about its performance in practice is called development work and forms a very important part of the designer's job.

When a new design or modification is received by the factory it is the job of the Production Planning Department to decide the best method of machining the parts, the machines which will be required and the time that it will take to complete the operation. In this way it is possible to keep the machine tools used efficiently and the cost of machining down to a minimum. Normally, the most accurate work takes longest to complete, uses the most expensive machine tools and the most skilled and therefore costly labour—factors which demonstrate the importance of keeping tolerances as large as possible.

If only one or two components are required to be made these can be produced individually by the machine operators from drawings. As the numbers increase it becomes worthwhile to make special equipment to drill holes or weld pieces together without having to measure and mark out positions each time. These devices are called jigs and fixtures.

I spent some time helping to design a few of these and it was not always as easy as it looked. I can recall one framework which I designed to locate parts to be welded together to make a cultivator frame. Fortunately, an experienced eye was cast over my effort before the drawings were released, which was just as well because once the cultivator frame had been welded it would have been impossible to separate it from the location framework!

There are several ways in which the machines and equipment can be laid out in the factory and this is often determined by the type of work undertaken. As products change from time to time so a change in layout might be required to ensure that the work flows as smoothly as possible. The Plant Layout and Work Study Departments were responsible for keeping a watchful eye on this aspect.

It was not apparent to me at first just how useful experience in these departments would be. A subsequent course on Work Study at College demonstrated just how universal the correct layout of the operators workplace is. These principles apply equally well to the machine tool operator and the combine harvester operator.

I regarded my week with the Safety Officer and Medical Officer as very important. Safety is basically a personal matter, the importance and relevance of which each individual has to judge for himself. Management must promote safety codes and standards but the key to accident prevention rests with the worker and his attitude of mind. On the farm in Kent I shall never forget the most experienced combine operator removing a guard and pulling a slipping vee belt by hand. I was standing next to him and distinctly remember remarking about the danger. Although the technique had worked satisfactorily several times before, on this occasion the belt gripped more quickly and crushed the unfortunate man's fingers between belt and pulley causing serious and permanent damage to his hand.

With so many parts being made and assembled so quickly,

some means of inspection and control of quality is essential in the factory. Material which is used for making the parts must be of a consistent standard and quality if failures in service are to be kept to a minimum. While I was working with this department a large number of gearboxes were rejected by the assembly line inspectors as being too noisy. Each gearbox was carefully dismantled and examined to determine why this was so.

Some evidence of damage to the gears was found but the problem was to discover how this occurred. The gears were machined from forged blanks supplied to the factory in batches. They received five machining operations and two heat treatment processes to harden the surface of the teeth. There was a lot of handling in between and it was thought that the trouble occurred during one of these handling operations. But where? Only several days of detection work revealed the cause. This was eventually overcome by changing the production method and lessening the amount of handling. A change which was only made after consultation and discussion with the design engineer.

When I was talking about becoming an agricultural engineer at school, several of my classmates would remark that it was little more than being a glorified blacksmith! These and other similar remarks made by my colleagues at Birmingham, who worked in the aircraft and machine tool industries, led me to believe that my chosen branch of engineering was relatively crude and imprecise. The high standards worked to by the quality control and inspection personnel soon helped me to correct this impression. Some of the tolerances specified on drawings were to within two and a half thousandths of a millimeter or one ten thousandth of an inch. This is an accuracy that even the aircraft industry does not meet every day.

One area where such standards of accuracy was required was the hydraulic pump control valve. The need for accuracy was so great that mating parts had to be matched by only assembling valves made near the upper tolerance limit with guides similarly made.

This procedure is not usually followed in practice. If a bracket snaps on the plough it is possible to obtain a spare part from a dealer which will fit—one hopes without trouble. This system of limits and fits as it is called, closely related to the principles of tolerances was first devised at the turn of the century by a man called Newall. He was making sheep shearing machines at the time which were selling very well in this country. The wool industry was just beginning to develop in Australia and sheep farmers began to order shearing machines from the U.K. All went well until parts began to wear and Newall received requests for replacements—which didn't fit! Tolerances are now arranged for mating parts so that no matter where the final dimension comes within the tolerance, the parts still go together successfully.

The nerve centre of the factory was the Production Control Department. The sales or marketing people supplied the office with orders and it was up to the production control staff to see that the right quantity and type of components were made at the right time to make tractors according to these orders. With only five tractor models in the range this would seem to be a simple problem. Variations in specification like different tyre sizes, exhaust systems and gear ratios produced over one thousand combinations which made things much more difficult.

To suggest that the assembly line be stopped or even slowed down because of some manufacturing or design problem ranked equal in importance and seriousness to announcing that the company strongroom was about to be burgled! Production Control was a vital function indeed.

A week in the Purchasing Department provided a useful insight into the financial implications of design and manufacture on this scale. The department had to arrange for the purchase and supply—at the right price—of all raw materials and parts which were bought out, such as batteries, instruments and wheels. The staff would sleep uneasily when there was a steel workers strike or a breakdown in the transport system due to bad weather.

The Company could not afford to purchase more than about three days' supply of 'fodder' for their 'hungry monster'. The cost of providing even this short reserve was several hundred thousand pounds.

I felt strongly at this time that a Design Engineer who had ambition to gain more responsibility and promotion leading to management positions really must learn as much about the financial operations in his Company as he can. Financial implications are so often the final arbiter in engineering problems. The designer who is able from experience to anticipate these implications is more likely to arrive at a successful solution in a shorter time than one who concentrates on the technical aspects alone.

There is little point in making a product in any quantity unless there are customers who are prepared to buy. Manufacture needs to be closely matched to sales requirements.

The company selling the product must be prepared to provide the customer with after-sales service. There must be some sort of guarantee to replace parts which fail early in the life of a machine. If the machine isn't working properly then the company must provide advice and assistance. Failure to do this can quickly lead potential new customers to look elsewhere for an alternative.

Making sales and providing service is the responsibility of Marketing. Closely associated with this work is Product Planning, the group of people who are thinking about the farmers' future requirements.

My training programme included visits to each of these departments. So many people are contributing to the total activity which is necessary to get a machine on to the farm. It is clear to see why an ability to work as a member of a team is considered to be an essential characteristic amongst those employed in industry.

8

PROFESSIONAL STATUS IN SIGHT: APPLIED RESEARCH

By now I was nearing the end of my practical training periods. For the last six months I planned to work more closely with the wide range of farm machinery produced by the company. With only one three month period remaining at College, there were many academic matters to be sorted out before the final examinations. My current objective was hopefully about to be achieved and with it, deeper entry into the agricultural engineering profession.

The Company Training School was set in magnificent surroundings close to Stoneleigh Abbey a few miles from Coventry. Its primary purpose was to provide training courses in the correct use, maintenance and repair of the Company's products for representatives of the dealer and distributor network. Being a large company with worldwide sales, several of the students at the School came from overseas. Not all the equipment used and demonstrated at the School was made in Britain. Several items such as rice harvesters and transplanters and huge disc ploughs were built in the company's overseas factories which did not enjoy the training facilities available at Stoneleigh. It was a strange sight to observe Malayan workers preparing rice paddy fields in the heart of the Warwickshire countryside!

The school organised a series of training courses covering the various types of equipment designed for mechanics, service engineers, or managers. I was fortunate enough to attend several of these courses during my stay. It was quite a change to return to the organised type of training after being so long touring the various factories. The courses were concentrated and well documented, which meant that I was able to build up a lot of experience in a short period of time. A feature not always present during my factory training.

My final three months at Birmingham simply flew past. According to the internal assessors my efforts were not to be in vain. Provided that I satisfactorily completed the Industrial Training Programme it seemed likely that I would obtain a Diploma in the upper second class. Good news indeed!

On my return to Coventry I was informed that my final training period would be spent working independently as a development engineer. This was worth more to me than a substantial pay rise. Although I had tried to work independently where possible during the earlier training periods, much of my work was conducted under close supervision. From the mistakes which I made the reason for this approach was obvious, although not always palatable. I felt I had reached this point on the farm when I first had sole charge of the drying machinery. Now once again, I had climbed a different ladder to a point where it was accepted that I no longer needed someone to hold my hand.

For some time the company had been working with manufacturers of artificial fertilizer to try and improve the method of applying it to the soil. The usual practice was either to dissolve it in water and spray it on, or to use a dispenser to apply it in powder or granule form, several rows at a time. The degree to which it could be concentrated was limited by the accuracy of the distributing mechanisms. If fertilizer is placed in direct contact with the seed disastrous effects can result. Water is not always readily available and in any event is bulky and costly to transport. Powder is difficult to handle and in moist atmospheres tends to bind together restricting the flow for distribution.

The problem I was given to investigate and solve, narrowed down to the development of an accurate dispensing mechanism which would enable concentrated fertilizer to be placed in the soil near seeds. I was to be entirely responsible for the programme and all decisions within the programme. One or two test engineers would be available to conduct any tests that I might require but they would be working to my instructions. An official report would be required at the end of the period.

Rather than try to start with a blank sheet of paper, I felt that the most appropriate first step would be to examine what had already been used and to form an opinion as to why they failed. Published test reports helped in this respect. In view of the limited time available, I would have been hard pressed to design, test and develop a mechanism in three months. Five years previously I would have probably tried to develop the whole machine from scratch and got nowhere in the process. One is always required to work fast and effectively in industry but only towards achievable targets. Attempting the impossible is of little value to anyone.

From my preliminary evaluation I decided to modify and further develop a device which had been tried out in Canada. Having prepared my own drawings, an experimental rig was made in the workshops. It was a great help when deciding on the shape and material for the various parts to think back over my own experiences in the workshop. In this way I was able to arrive at a simple and low cost device which could fulfil the purpose I had in mind.

The test programme was built up in stages. It seemed important to me that any test results I obtained should be repeatable. The problem with fertilizer granules was that their shape varied and it tended to be hygroscopic so its flow properties changed according to the relative humidity. I managed to overcome this problem by using small plastic cylinders which were about the same size as the fertilizer granules. Tests showed that a linear relationship existed between delivery of fertilizer and plastic cylinders provided new fertilizer was used under constant humidity conditions.

My tests were designed to start from the idealised state when the rig was stationary and level. Tilting backwards and forwards, then from side to side was used to determine the effect on accuracy of delivery. The whole rig was then run over bumps to simulate field conditions.

Although the device worked satisfactorily during the tests and within the performance specification set down, it never became incorporated in a production machine. Company plans and policies are continually changing and being modified to take account of current needs. As I had seen on several previous occasions there are a large number of carefully planned steps to be taken before a product reaches the customer and my idea together with the evidence that it worked only represented the first two or three of these steps.

This failure to put the results of my work into practice did not seem particularly important to me at the time. After all, it was only a three month exercise and much more work design and testing would need to be put in probably costing hundreds of pounds. The fertilizer manufacturers may have decided not to go ahead with their plans for a more concentrated, less bulky product. This would have removed the need for greater accuracy of distribution. I suppose that I could have taken a much different view and interpreted the reaction to my work as rejection of my services and left the Company shortly afterwards. I believed then and now that this would have been a very wrong move to make.

I wonder whether my reaction would have been the same in different circumstances? If this was just one of three or four of my recent projects which had all suffered the same fate, the drain on my morale may have produced an entirely different attitude to the decision makers in high places and consequently the Company. It is not difficult to appreciate how the frustration that is sometimes shown by designers and other engineers arises in this sort of situation. Perhaps this is just one of the tests which a professional person must face. Rather than allowing himself to be dragged down by setbacks he must learn quickly to recover and rise above them.

It would take a month or six weeks before the result of my Dip. Tech. course was confirmed. The Company decided to wait until this was available before making a formal offer of appointment. I did know, however, that when it was made I should be eventually working as a design engineer on tractors.

Problems are always arising in industry and people needed to help solve them. I was not to be allowed to remain idle for long as the Company's recently introduced potato harvester had developed an undesirably large number of transmission shaft failures. It was my job to help investigate the problem. Selected farms in Northern Ireland were chosen for the trials. As I discovered on arrival they were noted for their difficult conditions including heavy soil and many stones which tended to jam in the mechanism.

Working on some of the more remote farms I recalled my days on vehicle recovery in Kenya and Malaya. The teamwork necessary to get out to a machine which had failed, to put it right often in difficult circumstances, was in many ways similar. In this situation however, it was important to get the machines working again as quickly as possible so that the farmer and his team of six machine operators could get back to the task of gathering the potato harvest as quickly as possible.

A careful log was kept of the hours worked, conditions of operation and failures. In this way we were able to build up a picture of the circumstances which led to particular failures including the suspect transmission shaft. Weekly reports were flown back to Coventry to assist the design engineer in his solution of the problem. After a short time a stronger type of steel was used for the shaft and this seemed to do the trick.

The investigation completed, I returned to Coventry eager to take up my first appointment at professional level. To become a professional engineer it is necessary to do two things. One is to

APPLIED RESEARCH 87

obtain appropriate academic qualifications which I had now done. The other requirement, which I regard as being of equal importance, is to gain membership of a professional institution.

The latter course of action identifies the individual with a particular branch of engineering. Through meetings and publications he is kept up to date with the latest progress and developments in his subject. Employers often advertise vacancies in these publications so he is able to use the service of his institution when seeking new career experience or advancement. Several employers in fact, ask for membership of particular institutions when seeking new staff. At the meetings he has the opportunity of making himself known to people with similar professional interests either by presenting papers himself or taking part in the discussions.

It took some time for me to appreciate these reasons for the existence of professional institutions. I found it helpful to consider them in relation to something with which I was more familiar and more readily understood—professional football.

Membership of an institution represented playing in the Football League. The Division was related to grade of membership. Being kept up to date was equivalent to training sessions—maintaining a level of fitness in order to practice the profession.

Transfer to a bigger club and movement to a higher division would certainly be easier if the player was already in the League. The analogy also extended to the importance of being an effective member of a team.

There are several grades of membership designed to suit the needs of individuals at various stages of their career development. These include two grades with the self-explanatory titles of Student and Graduate. After a graduate has spent some years in industry including a specified period in a responsible position he becomes eligible for what are known as the Corporate grades. The use of the word Corporate implies full membership. There are two Corporate grades, the first of which is the Member grade. After further practice as a professional engineer rising to senior positions the Member can be elected to the highest grade which is Fellow.

It is important for the membership to keep in touch with those who, although not qualified academically, are still closely connected with the particular branch of engineering and making a valuable contribution. An additional grade called Associate exists for this purpose. Engineering institutions may differ slightly from this general description, but the overall pattern will remain the same.

With this view of professional institutions in mind I felt that it was necessary for me to join. The Institution of Agricultural Engineers was an obvious choice and I became a student member during my study for the N.D.Agr.E. which was an acceptable qualification for entry to the graduate grade.

It is not uncommon in our language for certain words to become overworked. The term 'engineer' is one of these words and can be used to describe a fifteen-year-old fitter's mate with one week's experience and the Chief Engineer of Rolls Royce. The objection to this state of affairs was not intended to belittle the work or contribution of the fitter's mate—after all Sir Henry Royce once described himself as 'just a mechanic'. Government and the senior institutions wanted to create a new title which would more accurately identify the professional engineer in order to establish his professional status and that of his profession. This involved setting up a Royal Charter which was conferred upon some thirteen institutions. Members of these institutions who reached one of the corporate grades were entitled to register as Chartered Engineers and could use the letters C.Eng. after their name.

There can be little doubt that this action had had the effect of improving the status of the engineer in society. Professional engineers can be readily identified both nationally and internationally. More people now recognise that engineering has an increasingly important part to play in our future.

The relevance of this to agricultural engineering is that its institution was not among those to be included in the Royal Charter. There was no doubt in my mind that I wanted to become a C.Eng. The only solution available then and now was to join two institutions! There is some hope that, in the future, it will be possible to register as a C.Eng. through the I.Agr.E.

I chose the Institution of Mechanical Engineers as being the Chartered Body nearest to my interests. Many professional agricultural engineers active in Field Engineering aspects choose the Institution of Civil Engineers.

My Dip. Tech. enabled me to qualify for graduate status of the I.Mech.E. It now remained for me to gain a few years' experience in 'responsible positions' before I was eligible to apply for corporate membership of each institution. Registration as a C.Eng. is only possible for corporate members of Chartered bodies so I would also have to wait a while for this title.

It was interesting to see the way in which the professional concept was reflected in the company organisation. My appointment to the staff of the Engineering Division was finally confirmed and described as 'Design Engineer 5'. This was the lowest grade just above the professional line. I received a detailed statement which described the type of responsibilities and duties that I may be given. Once again I found myself at the bottom of a ladder. I eagerly awaited my first assignment.

9

MAKING FARM MACHINES WORK:
AS A DESIGN ENGINEER

AT ONE STAGE I wondered whether I should ever have the chance of helping to design and develop tractors. It was still agreed that I should eventually work in this field, but I was first to join the Company Standards Engineer and two clerical assistants to tackle a more immediate task. A new range of combine harvesters was about to be introduced.

The intention was to sell the machines in many overseas countries as well as in the United Kingdom. Before doing this the Company wanted to make sure that the harvesters would work satisfactorily under the wide range of conditions which might be encountered. It was also necessary to get as much testing done in the shortest possible time.

Harvesting cereals in any particular area only lasts for about two weeks. The requirement was to test the machines for several months to discover and overcome any weaknesses before releasing them to farmers. The plan was to start a group of machines working in North Africa early in the year and to follow the harvest through Italy, France, Switzerland, Portugal, Belgium, England and Scotland.

The programme would be completed during the following winter months by machines working in New Zealand and South Africa. In this way it was possible to use the harvesters more or less continuously for a whole year! Several crops were harvested including wheat, barley, oats, grass seeds, rice and lupins.

The small team of which I was a member had to co-ordinate this complex network of field operations. I could not find any reference to this sort of work in my job specification and I very much doubt whether the Standards Engineer could either. His normal work was to see that design engineers were kept informed of company, national and international engineering standards.

This was just one of several occasions when I found it necessary to keep a flexible and open mind about the relevance and value of particular duties to both myself and my employer. It is essential in modern industry to obtain experience in breadth as well as depth. Here was a project of undoubted breadth having international implications. Although the technical content from an engineering point of view was limited there was plenty of opportunity to gain useful experience of administration, report writing and communications.

In the course of any career development the relationship between the employer and employee is essentially one of buying and selling. The employer is prepared to pay a salary according to the service he receives from his employee. Viewed in the other direction, the employee during his education, training, acceptance of responsibilities and gaining job experience is building up a background which together with his personal qualities, he sells to those who wish to benefit from them. We have also seen the ways in which membership of a professional institution can assist the sale.

It is not always desirable to sell to the highest bidder since several other factors such as job satisfaction are involved. The greatest rewards in terms of good financial return and a wide choice of position are normally available to those who make best use of their personal qualities in gaining a broad and thorough understanding and experience of their chosen profession.

Each fortnight a meeting was held to review progress which was attended by very senior members of the Company including representatives from the Scottish factory and occasionally from the United States and Canada. It was the responsibility of the co-ordinating team to prepare an up-to-date report of the situation for those attending the meeting. The reports were over a hundred pages long and were often printed at about 2.00 a.m. on the morning of the meeting. I imagine that this was as near as I ever came to experiencing something of the life and work of a newspaper editor!

I felt very important being in the midst of an activity which was receiving so much attention and obviously costing the Company tens of thousands of pounds. A few members of the Field Test team were permanently posted overseas to work with the machines. When a serious difficulty arose which, for some reason they could not tackle, the meeting decided to send a more senior man to investigate. Tomorrow would not be soon enough. The man was on a flight to the South of France later that afternoon.

The Standards Engineer and I were required to attend each meeting to present a verbal report and to provide any other information which might be required. All sorts of unusual problems had to be faced in the day to day running of the exercise. On one occasion three machines were dispatched from the factory for testing in southern Italy, only to be mysteriously lost for some ten days. After several inquiries they were eventually located in a railway siding somewhere in Yugoslavia!

There was a vast amount of data coming into the office each day. My knowledge of statistics proved very helpful when attempting to assemble all the figures in some meaningful form.

It was essential to maintain close contact with the harvester design team who were kept fully occupied providing the modifications necessary to overcome the problems which were arising as the test programme proceeded. One of the section leaders in this team was a lady who directed and supervised the work of several draughtsmen. Her technical training was obtained through the National Certificate course. She was very good at her job and her engineering ability was much respected by her colleagues. At this time it was quite unusual to find women in

MAKING FARM MACHINES WORK 93

engineering but more are now beginning to join the profession.
Once the European part of the operation was completed and machines had been sent to New Zealand and South Africa for the final stages of the test programme it was decided to disband the co-ordinating group. Each member returned to his former offices and I moved on to join the tractor design team.

It was about at this time that I first heard of the establishment of the National College of Agricultural Engineering. This was of considerable interest to me for two reasons. During my own education and training I had been very much aware of the lack of a degree level course in agricultural engineering. Having spent so much time listening to lectures both good and bad, I had developed an interest in the passing on of knowledge to others.

One of my colleagues was a part-time lecturer at a local technical college. While he was on vacation during term time I had taken one or two of his 'O' level maths classes. I enjoyed the experience which encouraged me to attend a technical teachers training course. The thought was beginning to grow in my mind that in a year or two when the work of the college began, I could well have something to contribute. This thought would have to be stored for the future since there was some tractor design work waiting to be tackled at the moment.

Tractor design was divided into two sections—modification and up-dating of existing models and new development. I was to work as a member of the second team helping to prepare designs for a new range of tractors.

Being an international company with its headquarters in North America, major programmes in Britain such as the new range of combine harvesters and also tractors were, initially at least, strongly influenced by events across the Atlantic. A similar range of tractors was to be produced by the North American company and much of their basic design work could be used as a starting point for the United Kingdom range. The ever present consideration was cost. Nevil Shute's definition was constantly brought to mind as I proceded to get involved in the provision of a new braking system and some aspects of gearbox design.

Designers of farm machinery particularly tractors are sometimes criticised for the slow rate at which design changes. The criticism may be justified but unfortunately innovation and change are not always compatible with economic viability.

The factory equipment and machine tools necessary to make a tractor for example cost millions of pounds. This cost must be recovered from the sale of tractors over a considerable period of time. The shorter this time period then obviously more of the cost must be charged to each tractor. Much of the manufacturing machinery is specialist and would not be readily adaptable to a very different type of tractor.

The best compromise is therefore, to achieve maximum development of design within the limitations of existing manufacturing techniques. I have heard people remark that most cars look alike nowadays. There are very good reasons why this has come about and here we have one of the most important ones.

Many of us would like to have a car which is different from the rest and this, thankfully, is still possible. The catch is we must be prepared to pay for it! The farmer is not buying farm machinery as a convenience or luxury item. Such equipment has become essential for his work and economic survival. He expects a sound product having quality and style but above all it must work for him as reliably and efficiently as possible and at the lowest cost.

It should be pointed out that manufacturing limitations only seriously apply to products made in large quantities. There are many types of farm machines designed and then made up in batches or relatively small numbers where innovation can be readily accepted—provided it can be justified commercially.

The need for creative thinking and original ideas in the design and development of all types of farm machinery and equipment is tremendous. In industry however, one should never forget that employers are in business basically to make money for their shareholders. If this simple fact is not recognised then survival in a competitive world becomes impractical.

In many minds there is a negative association with the word 'limitation'. It could be interpreted that what I have said about tractors leads to a dull and restrictive life for the designer. Well, this is one view which is held by a few people. After all there are some people who think negatively most of the time! It would be unrealistic if I were to suggest that I never felt the restrictive nature of the situation. The development of an ability to work successfully within these restrictions was recognised as a challenge and sometimes stimulated a solution which might not have materialised had there been a completely free choice.

Since I had previously learnt something about the co-ordination of test programmes, I was not surprised when given the job of conducting a similar exercise for the tractors. What I enjoyed most about this sort of work was the contact it gave me with people. It meant that for a time I could not continue with much design work myself, but this was more than compensated for by the contact I had with a much wider range of other people's designs.

One problem which often arises is that designers (including myself) tend to develop a mother/child relationship with their designs. This leads them to over-rate tests which have produced favourable results and to become quite protective when failures occur! It was not really a difficult task to maintain harmonious relations between designers, test engineers and the other specialists but one or two delicate moments did arise.

It is often said that in solving one problem you create another. This can happen in design. Farmers had asked for more power in the new range of tractors so it was decided to develop and fit larger engines. Being of the compression ignition type, noise levels increased slightly but what was more noticeable was the vibration of components in the vicinity of the operator.

Several of the field test engineers commented on this and design studies had to be made to find ways of reducing the effect. The main cause was the larger unbalanced forces in the engine being transmitted through the tractor to the operator's platform. Improved engine balance and one or two other design modifications overcame this problem but, as I was to discover in later years, we were only looking at the tip of the iceberg.

Several more serious and difficult problems were to emerge with respect to the effect of vibration and noise on operator health and efficiency for all types of tractor and other self-propelled machines—but more about that later.

The National College had by now been set up and after much careful thought and consideration I decided to apply to join the staff. My application was successful and it was with many regrets that I moved on from Massey-Ferguson but not away from industry. I say this because my new employers, the College, had been 'put in business' largely as a result of initiative from Industry and the I.Agr.E. The purpose was to provide a staff college for the industry which meant that close links between the two would need to be maintained.

A new chapter in my career was about to begin and I was very much looking forward to some experience on the other side of the bench.

10

THE OTHER SIDE OF THE BENCH: EDUCATING THE AGRICULTURAL ENGINEERS OF THE FUTURE

STANDING AT THE ENTRANCE to the College grounds one summer evening looking at the many buildings silhouetted against the skyline, there was an air of permanence and long standing about the site. Lights were shining from study bedrooms. Although it was getting late there was plenty of activity. In one room a student could be seen sitting at his desk surrounded by books, obviously intent upon mastering some principle or theory. Further along his floor there was a very different picture framed by the window. Eight or ten students were gathered for coffee enjoying a chat, a brief respite from the piles of lecture notes and handouts awaiting their return in a few minutes time. Undergraduate examinations were due to begin tomorrow morning. Successes were to be achieved and fates settled during the course of the next few days.

It is difficult to believe that in 1961 an excellent crop of potatoes had been gathered from fields on which the buildings now stood. During the following winter the site had been transformed by contractors' vehicles into a sea of mud and sticky clay dug up from the trenches in which foundations were to be laid. The College began to exist in the summer of 1962 using a single wooden hut almost marooned in the mud. That hut, together with a hastily erected Terrapin building had to perform all the functions necessary to run a College for the first few months. The Principal's office, staff studies, administration, lecture theatre, common room—I cannot quite remember about the toilets, but no doubt they were there as well.

During my interview for the job it was mentioned that there could be some building delays and because of this, conditions might be difficult for the first few weeks. In fact the weeks stretched to months and, whenever it rained hard, knee boots proved quite inadequate as the mud squelched over the top.

Fortunately there were only 15 students at the time to cater for—and an equal number of staff to attend to their needs! These had spent their first year at the College in a large country house in Essex loaned for the purpose by the Ford Motor Company. The second intake of students had been asked to delay their arrival until after Christmas when it was hoped that we would be in a better position to receive them. The New Year saw little improvement in the building situation. Sufficient space could be found for lectures on site but there was nowhere for our 50 students to eat and sleep.

Our well established neighbours, the National Institute of Agricultural Engineering, managed to squeeze about half the number into their hostel which was used for research workers and visitors. We had to book three or four rooms at a hotel on the edge of Woburn Abbey grounds. Those young men lived like lords for a few months. The remainder went to live in a zoo—in a house situated in the grounds of a private zoo to be more accurate! Their mid-morning coffee conversation was probably unique in the history of any college. 'Didn't fancy my toast this morning so I threw it to the vultures!' 'Did you manage to catch the wallaby last night after dinner?' 'What's that scratch on your arm?' 'You ought to get it seen to' 'Oh its nothing, that wretched eagle couldn't chose between my arm and the rabbit we were feeding it with at the weekend!'

One of the first buildings to be completed was the boiler house. Entirely below ground, the architects in a misguided flash of inspiration had decided that a large pond should be built on its

AGRICULTURAL ENGINEERS OF THE FUTURE

roof from which the enormous shining chimney protruded. Whether the design was intended to cushion the fall of those students who failed to reach the summit, or to permit the study of tropical fish farming techniques in a naturally heated pool, no one seems to know.

The first noticeable feature was that water began to cascade down the walls of the boiler house rising on the floor dangerously near the electrically controlled oil burners. In a short space of time the modern and shining boiler house would have become a slimy-walled, rust-filled grotto had action been delayed. An effective solution provided the builders with many headaches and sleepless nights. It was agreed that the College would man the bilge pumps. All hands were needed, this frequently included the Principal's during heavy rain. He became very knowledgeable about the boiler house during his many rescue operations!

The unusual and often difficult circumstances in which the College started its life at Silsoe helped considerably to establish a friendly and happy spirit of co-operation and teamwork. Students and staff alike, each had an important part to play and the sense of belonging was very strong. As the College has grown, the gum boots and sleeves rolled up at the ready approach has inevitably, but rightly disappeared. It is encouraging to see that so much of the original spirit remains.

The task of building up courses in Agricultural Engineering where none had previously existed, proved to be a once in a lifetime experience. The need had been identified by others, but it was now up to us to try and satisfy this need. I was often asking myself why I had made the change. From what I had seen as a student in industry there was a great difference between the life and work of engineers in teaching and industry. Perhaps the real challenge was to lessen this difference and hence the gap that exists between college and industry. I wanted to influence decisions about the use of drawing paper and give students a strong sense of cost when preparing designs.

There are many special problems that the agricultural engineer has to face which have made the profession particularly appealing to me. These must be effectively presented to students to prepare them for their later career. When designing a machine it is always important to take account of the environment and conditions under which the machine is to be used. For aircraft, runways and the air must be considered. Different types of road effect the design of cars. The equivalent in railway engineering is the track. How uniform and easily predictable these are compared with the soil in which many agricultural machines are expected to work! One has only to try and dig sand and then clay in gardens to notice the difference. Changing weather conditions can effect road, rail and air transport. The effect of weather on the soil can be equally as great, but often more complicated. It is a slow business to persuade crops to ripen uniformly for the benefit of machine designers. Even if this can be achieved the weather can sometimes create even greater problems by flattening crops as they stand in the field waiting to be harvested.

Thanks to the very low student/staff ratio my lecturing commitments during the first term were light. Like most of my newly appointed colleagues I had arrived straight from industry with virtually no teaching experience. It has always struck me as strange that it takes three years to train a schoolteacher, but for further education the only requirement is that you have a technical qualification and some practical experience in industry. I had at least stood up in front of a class before and found a lot of useful information in the brief teachers training course which I attended in Coventry.

Due to our tolerant and understanding students we were able gradually to learn how best to present information and explain techniques. I quickly found that the best way really to understand a concept is to try and teach it! No one likes being caught out on the other side of the bench, no matter how much they try to prepare themselves for it. I have always found that the best way to react in this sort of situation is honestly. Students questions provide a much more intensive examination of one's own knowledge and understanding than private study ever can.

A lot has been written elsewhere about the gap that exists between school and college. The belief that 'Sir' is as infallible as the concepts and theories he is teaching has to be overcome in both respects. The tutorial system which operates at Silsoe provides an opportunity for lecturer and student to meet and discuss topics in further detail. It is here that assumptions made for a particular theory can be examined more closely and their limitations considered in relation to practical problems. The lecturer may find during tutorial work that the discussion gets round to dealing with aspects of problems for which he is inadequately trained or prepared.

I believe that this situation is a healthy experience for the student. The way to a satisfactory solution can often be found by going to the appropriate textbook or research paper. In a subject which has considerable breadth and rapid rate of development like agricultural engineering, this approach is frequently necessary for the more advanced work.

Some students find that they do not come into sufficiently close contact with agricultural engineering proper at the beginning of the course. Provided that they can persist in applying themselves to the basic work and accept their Director of Studies' assurance that more interesting and applied work is just around the corner, then usually all is well. It is essential to build on sound foundations.

Many hours are spent in mathematics lectures getting to grips with differential equations. Later on these will be used to help solve many problems such as developing a control system for drying crops or feeding animals automatically. By careful selection of seeds it is possible to persuade plants to bear their fruit on tall bushes or short. Having a choice can make it very convenient when designing a machine to gather this fruit. Learning the language of differential equations and how to modify the growth of plants must come before putting the knowledge into practice.

What are Directors of Studies? Each student, once accepted for entry on to a course is allocated by the Principal to a member of staff who, during the student's college career—and beyond—is there to act as adviser and guide on all matters. These may range from fixing up a series of tutorials with the mathematics lecturer, to helping him sort out his girlfriend problems. Similarly girls among our students have been known to seek a Director's advice on boyfriend problems. The student/staff ratio has now risen to 7:1 but we still try to make each student feel that he is being treated as an individual. In a small college with less than 200 students it is possible to do this.

Being substantially residential, part of the College organisation includes a small group of staff called Wardens who live in the hostels traditionally to 'keep an eye' on the students. As Senior Warden since the College began to exist at Silsoe, this duty has added what I regard as an important and valued dimension to my work. Assisted by three of my academic colleagues we have tried to play down the traditional view, and replace it with a more constructive and positive system.

Students individually, and as a body have been given a generous measure of responsibility for the management of their own affairs. Generally, this approach has worked well and to the satisfaction of students and staff. The ability to work with people —to guide, persuade and direct in an acceptable fashion—is becoming of ever increasing importance in modern industry. By accepting responsibility and acting as representatives on committees and working parties, many students are able to develop this ability.

For the Wardens, living close to the students is not without its problems. I am sure the students would say the same! As the only married Warden, I at one time lived in a three bedroomed flat with my wife and children, on the ground floor of one of the hostels. The mutual difficulties such as children playing noisily when students wanted to study and students singing loudly when children wanted to sleep were overcome with remarkable ease. We all got along very well together as neighbours with just one exception. A few newly arrived students would get into the habit of calling at our flat during the evenings and weekends with quite minor requests like 'Do you know the times of buses to Bedford?' or 'Have you a road map of the area?' Naturally, we sought to discourage this traffic and our technique was for my wife to go to the door and, with an artificial but exceedingly formal tone, would enquire 'Is it urgent?' This approach had the desired effect but we were very glad that one recently arrived and very out of breath young man summoned sufficient courage to face my wife's demanding request and timidly reply 'Yes, please tell Mr. May that the Carpenter's hut is on fire'!

The disadvantage of being a small college in a rural situation needed to be recognised early on, and steps taken to overcome them if possible. One advantage claimed for a large university in which there are thousands of students studying many different subjects, is the opportunity for an exchange of views and ideas between the scientists, engineers, doctors and artists of tomorrow. This opportunity is, of course, very valuable but I wonder to what extent it actually takes place? In my own experience there is a strong tendency for social groups to form according to faculty or department. Mixed halls of residence help to break down this pattern but do not always succeed.

At Silsoe the very nature of the course provides no small part of the solution to this problem. Students are not confined in their course of study to one branch of the major engineering disciplines. In addition, botany, agriculture, economics and management all need to receive some attention.

The international flavour provided by an overseas student group representing some 20 per cent of all students, expands the interests and outlook of all members of College in another direction.

When late summer gives way to autumn we hardly notice the weakening sunshine, cold nights and early morning frosts. Our overseas students, often from a tropical climate, find the change in conditions quite a shock.

In return for advice about heavy sweaters and string vests from a well-known store, they soon begin to provide less travelled students with stories about their countries and traditions. 'Why I needed a camel in order to complete my practical training in the Sudan' or 'Life in Czechoslovakia today' no longer has to be read in the press or magazines. It can be heard first hand over coffee or a glass of beer!

Much to their credit, the students have formed close social and sporting links with several other colleges in the area. These colleges are concerned with Education, Physical Education, Aeronautics and Agriculture. Despite being the youngest of the colleges, our students have often taken the lead in organising and co-ordinating events and activities of the group which include dances, debates and social evenings.

Quite contrary to the fear that Silsoe might become just another field of cabbages in the Bedfordshire countryside, the extra mural activities of the College have flourished and a wide stimulating experience is available for all who seek it.

Sir Barnes Wallis, a designer extraordinary of airships, of dam buster and swing wing aircraft fame, once stated that the two most important personal requirements for successful design were enthusiasm and experience. The objective at Silsoe is to generate enthusiasm and provide experience. I am proud to be associated with this objective.

II

ACCEPTING THE CHALLENGE: RESEARCH AND DEVELOPMENT PROJECTS

THROUGHOUT ITS SHORT HISTORY the College has placed equal emphasis upon agricultural engineering at home and overseas. This has been necessary partly because increasing numbers of students from overseas have been coming to study at Silsoe. As a small country we depend a great deal upon our export trade. If agricultural machinery and equipment is to be developed for use overseas then up-to-date knowledge and experience of crops and conditions are required.

Although I no longer work in industry, there are many similarities between the outward looking approach at Silsoe and that adopted by industry. Not the least of these similarities is the need for hard work! I had heard about the dangers of stagnation in the educational world. There has been little time for this when trying to get a new college established, building up courses, and an image, preparing students to take part in a young profession, assisting that profession and the College to gain a sound reputation and wide acceptance.

The challenge continues with involvement in the problems of mechanising agriculture particularly in the developing countries. One of the most important ways in which staff keep in touch with these problems is by spending periods in the countries actually helping to solve them. On their return, the experience accumulated can be passed on to students in lectures. The future of agricultural engineering in the United Kingdom, and consequently of the College depends very much on an outward looking approach.

The effect of any inconvenience caused by temporary staff absences is minimised by the appointment of additional staff when people are away for a year or more.

At the moment College staff are working in Thailand, Malaysia, Sierra Leone, Nigeria, Uganda and Barbados. The visits to Thailand and Sierra Leone are mainly educational where, for a period of about two years, courses are being established or developed using experience acquired at Silsoe and in previous similar appointments. A recent major research project at Silsoe was concerned with developing a rice harvester based on a new operating principle. Instead of cutting the whole plant and passing it through the machine only the heads are threshed by the machine while the remainder of each plant is left standing in the field. The manufacturer who sponsored the project is now testing the machine in Malaysia and we have a member of our staff present as adviser.

The mechanisation of crops in Barbados particularly sugar cane needs careful planning and development. A mechanisation specialist is there to help ensure that all goes well.

Our agriculturalist is working with a team in Uganda to assist the selection of crops and varieties best suited to local conditions. Nigerians are beginning to run their own technical certificate courses in agricultural engineering. A two-week visit is being made by a member of staff to assist with this work. Last year he was in Korea and South Vietnam with a trade mission.

A Civil Engineering specialist has recently returned from Iran where he was working for a firm of consultants, and an Economist has just come back from a three-month advisory visit to Ceylon where thoughts are being given to local manufacture of farm equipment. In the past eighteen months this member of staff has also visited Peru and Panama advising on mechanisation programmes.

These examples serve to illustrate the outward looking approach of the College and the extent of the challenge facing experienced agricultural engineers. One or two of our graduates are beginning to take up similar appointments with firms of consultants and overseas governments. As more become sufficiently experienced the numbers will increase.

My most recent visit was to Uganda. In common with many developing countries particularly in Africa, Uganda has found that to import large quantities of machinery and equipment designed for use in Europe and North America sometimes creates more problems than it solves. Equipment may remain idle because of lack of trained operators or sufficiently skilled service mechanics. Farmers may have only very small areas of land to work and cannot afford their own machinery. Hire schemes have been tried but with only limited success. Farms are often many miles apart and someone has to pay for the time spent in transport between farms.

The most serious barrier to mechanisation in developing countries is probably lack of money. British manufacturers are very doubtful whether the peasant farmer will be able to afford his own equipment in the foreseeable future. The reaction is similar in other parts of the developed world. Before producing a new machine designed to suit the needs of developing countries, a manufacturer would want to know the 'size of the market'. He would expect to sell at least thousands, more likely tens of thousands and in some cases hundreds of thousands before considering the exercise justified.

Accurate assessment of the market potential is difficult. Conditions and requirements can vary widely between one country and the next. A challenge indeed! But what is the answer? There are plenty of statistics to show how serious the food shortage is in many parts of the world; how crude and inefficient are the traditional methods being used.

Several countries are beginning to see what they can do about the problem themselves. The Agricultural Engineering Department of the University in Uganda has developed a small tractor during the past ten years or so. The object is to make as much of the tractor as they can themselves, buying the complex components like gearboxes, wheels and tyres from the developed parts of the world. In this way it is hoped to arrive at a low cost, simple and appropriately sized power unit for the Ugandan Farmer. Several items of equipment have been built for use with the tractor. Once the tractor and equipment have been sufficiently developed for manufacture the plan is to build a factory to make them in.

The reason for my visit was to study the project for a short period and then advise the University on the next steps. There were many aspects to consider and I spent some time visiting the existing manufacturing facilities in the country and several farms to get an indication of the use to which the equipment would be put. A few large tractors were working successfully, oxen were common in some areas while in others the land was cultivated using traditional hoes called *jembes*. The small tractor was intended to work alongside these existing methods, gradually replacing ox cultivation and some hand work. Unlike people, Nature refuses to be hurried where revolution is concerned.

There were moments for relaxation including a magnificent boat trip along the Blue Nile which is littered with playful hippo and infested with crocodile which were often to be seen stretched out on the banks in a deceptively lazy fashion.

It was agreed that Silsoe should add the finishing touches to the design and test the tractor under British conditions. The latest prototype was flown back from Uganda for closer examination. I'm sure it would have been easier to get half a dozen of the crocodiles back than that tractor! What the Customs officials just could not understand was why the world's largest exporter of agricultural tractors had suddenly taken it into its head to start importing from Uganda.

The development work is now almost complete and next year we plan to send another member of staff to help set up the factory. One of our recent graduates has already started work on the factory in Uganda.

As in most countries of the world, interest in agricultural engineering in the Caribbean is mounting. In 1970, an international seminar was held at the University of West Indies in Trinidad to receive papers presented by delegates from many countries. I was invited to represent the United Kingdom and was whisked off in a VC10 jet arriving at the Port of Spain airstrip only eight hours after leaving Heathrow. The temperature had risen from 0° C to 32° C. I was getting the experience of postgraduates arriving at Silsoe but in reverse! I had also gained six hours into the bargain. Luckily the excitement of arrival in a new part of the world was enough to carry me through a cocktail party which must have gone on until 7.00 a.m. British time.

We were given an extremely friendly welcome and although the activities of the day were demanding there was plenty of air conditioning in the buildings. The heat became quite pleasant during the evening when the temperature would drop to 28 or 29° C and a gentle breeze blew in from the sea. Each evening there would be a social gathering of some form. As one might imagine the fruits were quite exotic and the rum very palatable!

The M.C.C. touring party were on the island at the time. There was so much to see and do that I had to decline an invitation to attend the Consulate dinner party given in their honour and also missed the match the following day. It is little wonder that the West Indians can field such an excellent cricket team when the only alternative I saw to playing in a steel band was wielding a cricket bat! Literally hundreds of boys seemed to be playing on makeshift pitches all over the island. The weather of course, was very much in their favour.

Several visits were made to agricultural projects. The rapidly growing tourist industry depended for much of its horticultural produce on Florida. This was naturally an expensive operation and one large scheme of about 3,000 acres was intended to encourage local farmers to produce cucumbers and other crops themselves. A river had been diverted to irrigate the land and mechanical aids were available for hire on a co-operative basis. Some were achieving high yields and good profits, but others, although cheerful and smiling (the national characteristic), were finding it very difficult to cope with complexities of machines, fertilizers, and seed varieties. The concept was good, the potential great, but the need for advisers and guidance was even greater.

The one word next to cricket which first comes to my mind when thinking about my trip to the Caribbean was sugar. The cane stood proudly erect mile after mile. A huge processing plant was operating at full tilt to produce the raw brown crystals for shipment to Britain. Mechanisation of the harvest is progressing slowly but is inseparably linked to a serious social problem—unemployment. Many workers who at present cut cane by hand, fear that machines will create mass unemployment on the island. In Lancashire, workers burnt the looms. In Trinidad the technique is to burn the cane. This makes it unfit for mechanical harvesting and keeps the hand cutters in work.

While the need for agricultural engineers is increasing throughout the world so is the need for an ability to take account of all factors associated with a particular problem. This undoubtedly increases both the interest and challenge for agricultural engineers of the future.

Whatever the outcome of the Common Market negotiations, the relationships between agricultural engineers in Europe are likely to strengthen during the new few years. I was therefore glad of the opportunity to make a six-week tour of research, public testing and educational centres in Holland, Germany, Denmark, Sweden and Norway soon after joining the College.

In addition to the valuable technical experience acquired,

contacts were made with several workers in agricultural engineering most of whom spoke English perfectly. I wonder how many British engineers speak one or more foreign languages? As world travel becomes more common so the need to communicate in more than one language will increase.

When I am not lecturing, giving tutorials, acting as Senior Warden, or galloping around the World, what do I do with my spare time? For as long as I can remember, I have always been interested in people. Since joining the College it has gradually occurred to me that I might make myself useful by combining this interest in people with my engineering career in a professional sense. The outcome has been for me to spend whatever time I can find to study the operator of farm machinery. All too often in the past, machines have been designed according to well established engineering principles and the operator added almost as an afterthought like the snowman on a well iced Christmas Cake.

A relatively new subject area called ergonomics (from the Greek *ergo*—work, *nomos*—law) is beginning to be applied to agricultural engineering. At Silsoe we are working mainly on simple systems such as crank operated maize shellers and knapsack sprayers. Other establishments, such as the National Institute of Agricultural Engineering, our next-door neighbour, concentrate on the more complex machines such as tractors. We keep in close touch with this work at the College and occasionally take part in the Institute programme.

The work for which I am responsible recognises that, particularly in the developing countries, farmers will be concentrating on the use of simple, often hand operated machinery for some time to come. By measuring the amount of energy man uses when operating these machines, perhaps in different ways, it is possible to improve the design either to produce a greater output for the same human input or to reduce the input energy requirements if these are excessive. As a result of modifications carried out on the maize sheller to enable the operator to pedal instead of crank, almost 40 per cent reduction in energy requirement was measured.

Our links with several universities and research establishments overseas keeps us informed about problems which require attention and some joint projects are undertaken with similar work going on in temperate and tropical climates for comparison purposes.

Those machines which are becoming more complex in design often tend to be more difficult to operate. Mental as well as physical stress needs to be considered as this trend develops.

At the Institute, work is being done to seek ways of overcoming the serious environmental problems affecting the operators of farm machinery. Medical authorities have found that it is necessary to keep the noise level close to the operator's ear below 90 decibels if damage to the ear is to be avoided. Many tractors in current production produce noise levels around this figure. When the flexible weather cabs are fitted, the level increases to 90–95 decibels. Metal cabs are becoming common now that safety frames are a legal requirement on all new tractors. For these cabs the noise level is in the range 95–100 dB.

When I had presented my paper in Trinidad (it was on the subject of operator efficiency and had referred to the noise problem), a member of the audience came over to me saying that I had answered a question which had been on his mind for some time. He never could understand why a friend of his who operated crawler tractors on the sugar plantations always had his television set at full volume which almost made his little house shake at the foundations—now he knew.

What can be done about this—and other increasingly serious environmental problems. Hand held chain saws produce vibration and noise which have harmful effects on health. The incidence of spinal injuries amongst several hundred tractor drivers was found to be five times higher than for the general population. Care in the use of toxic chemicals as herbicides or insecticides is obvious to most people but the build up of carbon monoxide levels in glasshouses is insidious.

My old company are honest enough to supply a set of ear muffs with every tractor sold and suspension seats are available. Much remains to be done to protect the operator's health and increase his efficiency. This is just another aspect of the enormous and absorbing challenge which the agricultural engineer of tomorrow must be prepared to accept.

12

THE PROOF OF THE PUDDING

You may by now be wondering how recent graduate entrants to the profession are getting on. What is the background of these people? How did they find life and work at College? Did they get interesting jobs or further training?

Some of the earlier Silsoe graduates have outlined their own impressions and experiences to try and provide answers to these questions.

David joined the College in 1963 after spending a year gaining practical experience on his father's farm in Hampshire. He had four 'A' levels in Maths and Science subjects, two more than the minimum requirement. He writes:—

> 'Everyone hears something when they are young which seems to stick in their mind. I can remember my father, a farmer, once telling me "Industries may come and go but people will always need food". All the same, my youthful ambition was to become an airline pilot—I think that the glamour appealed to me. On being turned down by the one airline that I approached, fortunately (unfortunately I thought then), I remembered my father's words and decided to turn to agriculture as a career. Now that the world seems to be heading towards another recession, I consider myself very fortunate to be working in the essential industry of Agricultural Engineering.
>
> 'After getting a few 'A' levels at school, I started working for my father, but I soon found that I much preferred to work with spanners and machines rather than with udder cloths and cows, so that when I found myself faced with the

choice of going to an agricultural college, or the National College of Agricultural Engineering, I turned to the engineering side of agriculture. At Silsoe I was at first surprised by the number of subjects being taught: I hadn't realised that agricultural engineering included such topics as Field Engineering and Environmental Control, besides the Design of Agricultural Machines.

'I soon adjusted to the new academic and social way of life and settled down to the three-year course, working hard and playing hard. Unfortunately, I was better at playing than working and so I ended up with a Third instead of a Second or First, but I managed to obtain a Graduate Apprenticeship with a company that manufactures grassland machinery.

'It wasn't too long before I was working as a design draughtsman on new machinery. I found that the academic training I had received at College helped me analyse and solve the problems that I was meeting in my work. A mind that has undergone what I call the academic stretching process seems to produce more original ideas and solutions than a person who has been trained on a practical experience basis.

'An opportunity came along for me to work on more academic problems than I had been working on previously. This was in the field of Standards Engineering. I was soon at grips with such problems as Procedures, Standardisation, Metrication and so on, but I think that perhaps it would be better called Political or Financial Engineering, since most problems have political or financial overtones. With this work I am helping the company to save money which, in turn, helps to keep down the cost of the machines to the farmer.

'I am very happy in my work—it stimulates the mind and I feel that I am aiding World Agriculture in my own small way. I would like to do more.'

There seem to be several similarities between David's career and my own. Standards are such an important part of design that he could be regarded as still working in the design field—gaining valuable experience for the future. Since he lives and works within easy travelling distance from Silsoe, David often returns to attend Branch Meetings of the Institution which are held at the College.

I always felt that Jim had set his sights firmly on becoming a design engineer but on the rare occasions when I see him he is either just back from a Marketing trip to Algeria and Morocco, or waiting for his visa to visit East Germany—but let's hear what he has to say:—

'The fact that my grandfather was a mechanical engineer and father a farmer provided me with the right background for agricultural engineering. In spite of that, I had little idea of what I wanted to do on leaving school. So, being mechanically minded, I applied for courses in civil, mechanical and agricultural engineering, hoping that the decision as to which course to follow would be made for me. Unfortunately, this was not the case, so I eventually chose agricultural engineering, knowing a little more about that than the others.

'After graduating from the N.C.A.E. in 1965, I joined a large international farm machinery manufacturing company as a graduate trainee. My immediate ambition was to become a farm machinery design engineer. However, to my dismay, I discovered during my training period that engineering in industry didn't come up to my expectations, and was not for me. On realising this, a major decision was necessary—should I stay with my present company in a non-engineering capacity, or move to a smaller company perhaps providing a different engineering environment. I chose the former, and accepted a position in U.K. marketing in the product planning section.

'On reflection, I think this has proved to be the correct decision. From the product side in U.K. marketing, I have now progressed to export marketing. This provides interesting work with a more direct involvement with the commercial aspects of marketing and also some overseas travel.

'Although my present position is far removed from my original ambition to be a design engineer, nevertheless, it is providing me with a worthwhile and interesting career.'

Like Jim, Martyn was a prominent sportsman at College. It was a great day for me on the cricket field if I could master Jim's bowling or manage to get Martyn out! From Canada, Martyn contributes the following:—

'Recently while completing an application form, I was asked the question, "State the skills you have acquired from other employment which can be related to the position for which you are applying." This question set my mind back to the time in 1962 when I finally decided to study agricultural engineering. Although my father was a physicist, I had been able to work on neighbourhood farms for a number of years. This work gave me a knowledge and love of agriculture which I still have today. My father's advice regarding the economic uncertainties of farming and my leanings toward engineering led me to Silsoe where the then recently formed National College of Agricultural Engineering was situated.

'When I started at Silsoe in January 1964 the College was still being built and we spent much of our first year dodging wheelbarrows and bricks. Indeed the school year did not begin until January because the residences were not complete. I spent September to December working as an apprentice mechanic at a large equipment dealers.

'Looking back over the course I feel that it achieved a

great deal. In the first two years the fundamentals were taught and in the third year the practical application of these fundamentals were considered. All through my time at Silsoe we were encouraged to think and develop as individuals.

'I have always felt a sense of belonging to the College—its size, at that time, not more than a hundred students, leads to a community spirit and pride unrivalled by larger colleges and universities which I subsequently attended. In order to keep the many sports clubs viable members often played on more than one College team. I was fortunate enough to play on the rugby, hockey and cricket teams, being secretary to both the rugby and cricket clubs.

'During my final year at Silsoe, I attended all the careers forums arranged by the Careers Office. After a number of interviews I still could not decide which career would suit me, so I eventually decided to join the Voluntary Service Overseas Organisation and was sent as an Agricultural Engineer to Uganda.

'My year in Uganda proved to be a wonderful learning experience and I have never regretted the time spent there. I had been offered a chance to do graduate work in Canada at the conclusion of my year in Uganda. I spent nearly eighteen months study for a Master of Science degree in hydrology at the University of Guelph, in Ontario. On completion of my M.Sc. degree I joined the Ontario Department of Agriculture and Food as an Agricultural Engineering Extension Specialist (similar to the work of engineers of the Agricultural Development Advisory Services in U.K.).

'During my time with O.D.A.F. I became very interested in the effect of agriculture on the environment, and with the particular problem of animal waste treatment. I felt that my background in agricultural engineering and hydrology was not sufficient to allow me to work effectively on the pollution problem as applied to agriculture, thus I decided to go back to school once again!

'In September 1970 I registered in the Department of Chemical Engineering, McMaster University to study water pollution control, water and waste-water treatment. The majority of the work deals with municipal and industrial waste treatment, however I feel that I will eventually be able to move back into agricultural engineering and the waste treatment knowledge will make a useful contribution to helping solve the pollution problem as it affects agriculture.'

We have to go to Australia to find Terence who, for the present, has decided to pass on some of his knowledge and experience in a teaching post.

'After failing in Physics at a large London College, I had to rethink. Although father runs a small agricultural engineering business, agricultural engineering was not an immediate choice. Having carefully read the prospectus I decided to apply to the National College. Here I found that, as promised, the subject matter was varied and interesting, as well as being related to practical problems of a fundamental nature. The relatively small number of students and the close contact with the staff were important factors in making the years at N.C.A.E. significant. Many late night discussions over coffee centred on the world food problem and our contribution to its solution.

'For me the next step was indicated by a trip to the United States as a result of a scholarship that I was fortunate to receive in the vacation before my final year in 1967. In America my curiosity was aroused, and I wished to return for a longer period. My limited exposure to the American universities showed me that for my particular interest, soil and water engineering, there was more knowledge to be gained. Accordingly, after graduation at the National College I went to America to begin my Master's degree programme at Michigan State University.

'I spent two years at M.S.U. studying, working as a research assistant, and performing my own research on soil particle transport phenomena. As well as the advanced knowledge gained I was exposed to life in a "foreign" country. For America is foreign; the life, style and scale of values are different, and it all took some getting used to. Living in a large university community was a dynamic experience. I was elected President of the graduate students' organisation in the hall of residence on campus, and got to know something of the jigsaw puzzle of academic staff, students, and technical service personnel. During this time I met and married my wife, an American.

'After graduating, my wife and I returned to England to visit my parents and to look for a job. The urge to travel was strong, and after some searching, the field narrowed to positions in Africa and Australia. In the end Australia won, and I accepted a teaching position in a College of Advanced Education in Queensland. This college offers one of the two professional agricultural engineering courses in Australia. The students, although small in number, are enthusiastic. I find teaching an absorbing and stimulating task, and I am presently studying for my Certificate in Tertiary Teaching. A Ph.D. programme is a future possiblity, although I feel that I would like to expand my practical experience before making the academic life my career.'

Alasdair was in the same year as Terence and also decided to 'top up' his further education in the U.S.A. Alasdair writes:—

'On leaving school I had no definite career objectives. I considered a number of alternatives, and finally chose farming because I enjoyed working in the open air, and I was attracted by the variety farming offered. Not long after beginning a year's practical training in preparation for a National Diploma in Agriculture course I realised that I was more interested in the application of machines to agriculture, than in agriculture itself.

'About that time I heard that training was available in agricultural engineering, and I was fortunate to be able to complete the 'A' level requirements for entry to the National College by day-release at a local technical college.

'The College course work was interesting, and gave me a good basic training in pure and applied subjects. Theory was kept in perspective by practical experiments illustrating the difficulties in applying theory to real life problems.

'When I left College I went to the U.S.A. to gain some research experience and to travel. There I became interested in the use of computer techniques as an aid to the design of large agricultural systems. Now back in England I am working on a project to try to improve the efficiency of harvesting methods by the application of these techniques.

'This is a fast developing area which I find very challenging. Although there are many problems, I feel that it will only be a matter of time before the impact of computer techniques will be felt.

'One final point, because agricultural engineering is a broad field, it is possible to change jobs as one's interests and aptitudes develop, without wasting your training.'

I believe that Alasdair's final point is a very important one. If there was more space we might have asked another Jim to write something about his marketing experiences with Aston Martin in several countries, or Chris who went on to Cambridge to obtain a Ph.D. for research into the development of control systems for use in the steel producing industry.

13

IF I COULD START AGAIN

LOOKING BACK OVER MY CAREER so far, it might seem rather ordinary and even complacent, to reflect that I wouldn't have changed very much. The temptation to think about what might have been is there but given that the same sort of opportunities had presented themselves, I think that I would have taken much the same course of action. Sport would still have played a dominant part in my early life and no amount of counsel from my elders be they parents, masters or just friends of the family, would have made me change that for the 'more useful and rewarding pursuits'.

This sort of comment always seemed so dull to me as a schoolboy since it is only as one's own experience widens that the value of study becomes apparent. The concept of forty or so years of work in one job in one area appalled me at the time but it all seemed too far away to be worth bothering about. I sometimes really wonder whether I would have been a better or more successful person by getting down to my studies much earlier than I did.

During the first few years after leaving school I can recall quite clearly that several of my contemporaries who had all achieved better academic standards than myself, seemed to be destined for sound but ordinary careers. Perhaps I wasn't considered appropriate material for these 'office type' careers—at least, no one ever suggested that I might consider them; I wasn't really interested in any case. They all seemed to be too demanding

IF I COULD START AGAIN 123

too early. They all meant working indoors—at a desk. I wasn't looking for a clean job, necessarily, and the outdoor open air element had to feature quite prominently.

The other disadvantage with this sort of work concerned the early commitment necessary. Steady wages, pension after long service, promotion by length of service—the 'dead mans shoes' approach we used to call it. I am probably doing a great injustice to careers which many have found both satisfying and rewarding but would never have attracted me.

If I have any serious regret it would be the delay in reaching university-level education. At the time of leaving school the possibility was not even in my mind but I am sure that my formal education would have been achieved more rapidly and efficiently perhaps to a more advanced level had I followed a more conventional route. My regret ends at this point because there is another approach to the situation which the passage of time had shown me to be of merit.

Most students entering the degree courses at Silsoe for example do so straight after leaving school. This is the conventional thing to do and I would agree is in many cases the best course of action to take. On several occasions, however, the student finds difficulties in adapting to College life. School and home provide the foundations, and very important they are, but for some it is an inadequate preparation for what is to follow. It can be a sheltered and restrictive existence ill-suited to the freedom of thought and action experienced at College.

Passing directly from home, to school and then to college forms the first half of what I choose to call the 'closed loop' problem. It is possible for the individual to trace out this part of the loop without really moving outside the influence of the three centres; home, school and college. The system in fact encourages the individual to remain well within the spheres of influence. In order to do well at school, sport, societies, hobbies and trips abroad may and often do suffer. College work particularly for technologists is demanding to all except the brilliant. Because of this, sport declines and other social interests are restricted. Even members of the opposite sex receive only the most superficial attention!

The products of such a system frequently find themselves proceeding rapidly but surely around the second half of the loop either back to school as teachers or back to college as lecturers or research workers. Those who escape may do so only partially. There is little difference between conditions for research in some university departments and that which is carried out in many of the larger firms.

I do not question the quality and relevance of this research but the quantity is of doubtful value. The current and future need is for qualified men and women to interpret results of research and apply the results in industry. In order to be able to do this, much wider experience is required than that ordinarily obtained in the closed loop system. This experience includes understanding and working with people of widely differing backgrounds and operating as a member of a team perhaps as a leader, in all sorts of situations. I have found great benefit in gaining this type of experience at all stages of my school and college career rather than in one lump at the end. Although this was not planned in any way—it just happened—I would certainly try whenever possible to gain a wide experience of people and living . . . if I could start again.

It is already apparent in my fresh start that technology is attracting me once again. It is attracting me because it is concerned with people, machines and getting things done. I am assuming that an interest in science developed at school and that I still have to take account of an affinity for farming and outdoor living.

Hardly a pause is necessary at this point before drawing the conclusion that agricultural engineering is the career which I would choose again. But why? It isn't because I am not aware of other possibilities. My own work in industry has taken me close to and sometimes within the boundaries of mechanical and production engineering. At the National College I have had the opportunity to meet other engineers including civil and electrical, all of whom have significant contributions to make to the education and training of agricultural engineers. Architects, agriculturists, botanists, mathematicians and economists also work on the staff at the college and it is probably this wide range of interests that are relevant to agricultural engineering which makes the subject so absorbing and challenging for me.

In practice, it must be only rarely that young people actually choose a career that they are really sure about. Choosing implies making a decision in a short period of time. I think that I was as sure as most about what I wanted to do which was to combine an interest in farming with a faint ability in the physical sciences. It took me several years, however, to get around to being definite about agricultural engineering. Should it be farming or should it be engineering? If engineering then what sort of engineering? At the time I probably did not recognise that more than one form of engineering existed!

Choosing a career in my opinion is a lengthy and complex business which can be influenced a great deal by circumstances, luck, opportunities and the ability of the individual to recognise those opportunities and to take advantage of them. When I joined Massey-Ferguson as a trainee draughtsman several members of the engineering staff were convinced that I must be related to the Chief Engineer since no previous appointment had ever been made in this way before. I found the story amusing at the time and, while I had never met the Chief Engineer in my life before the first interview, I did not take any other action to check the rumour.

If approached in the right manner I have always found people to be most helpful. I do sometimes wonder, however, just how much my advisers were influenced by the thought that I was a member of the Chief Engineer's family! Would I have made less progress had the circumstances been different? Possibly not, but luck was definitely on my side in this case. An additional opportunity was presented, recognised and accepted.

Probably the greatest benefit of this slightly fraudulent situation was the confidence that it gave me in my first real encounter with industry and the people working in industry. It is most unlikely that this sort of opportunity would present itself again but I firmly believe that different opportunities of equal value would arise. They cannot be predicted and therefore could not be included in any plans that one may have. The important thing to remember is to recognise that they are there, to see them clearly, selecting the relevant ones, and be influenced by them just as much as is necessary to achieve progress in a chosen direction.

Whenever friends or acquaintances remark that they would have done something quite different with their lives if only they had had the right opportunities, I wonder whether it really is a lack of opportunity or lack of recognition and acceptance of opportunities.

Diagram 2 shown opposite gives a summary of the educational opportunities that exist today in agricultural engineering. General education to 'O' level standard is represented by a small circle and each additional concentric circle of larger radius represents a further year's academic study. Practical training has had to be left out of the diagram in order to keep it simple and easily understood.

Qualifications recognised as leading to corporate (full) membership of the Institution of Agricultural Engineers are shown as black dots and other qualifications held by those working in the agricultural engineering industry, as open dots. Qualifications in the field of agriculture are shown in the sector inclined to the left and those in non-agricultural branches of engineering in the sector inclined to the right. The qualifications within the cross-hatched area are recognised as leading to Chartered Engineer

IF I COULD START AGAIN

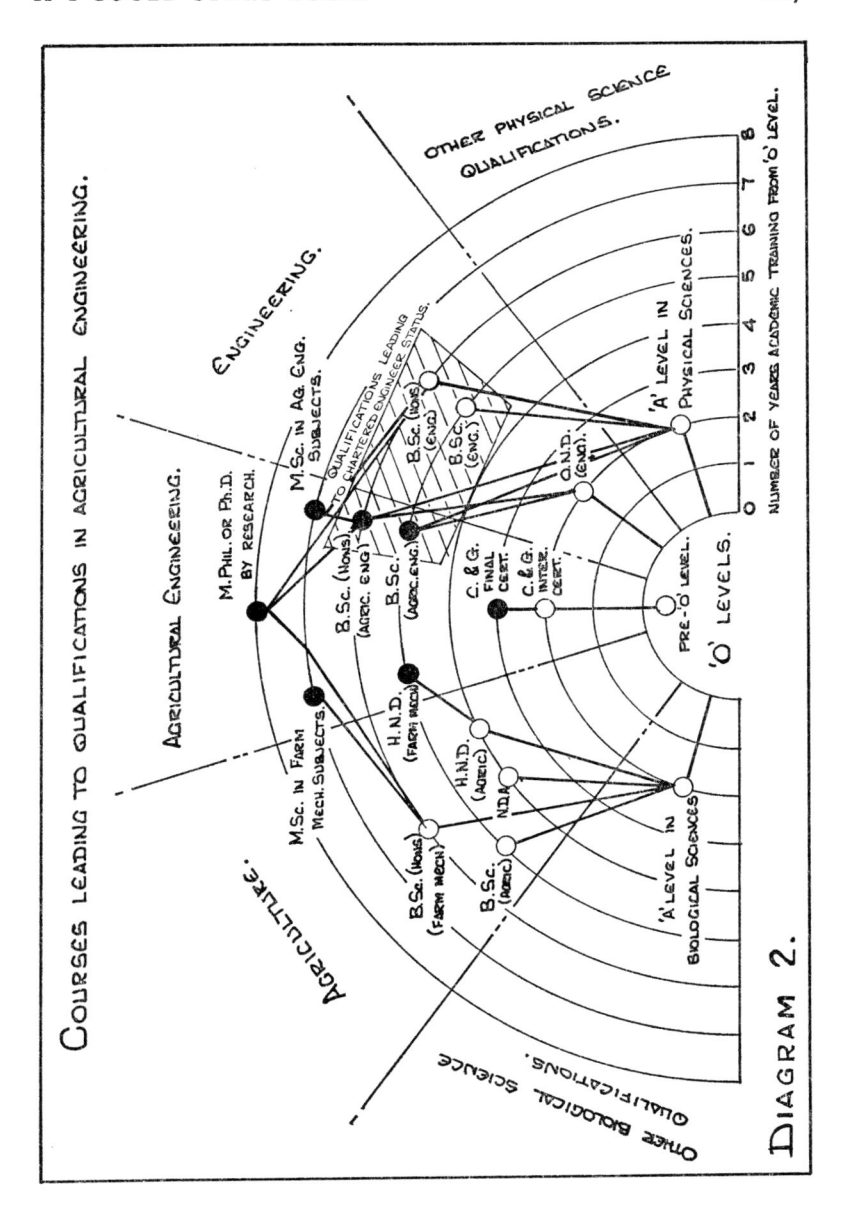

DIAGRAM 2.

Courses leading to qualifications in agricultural engineering.

status and include first degrees in both agricultural and other branches of engineering.

New courses have recently been introduced by the City and Guilds of London Institute to meet the need for more recruits at higher technician level. The first examinations in these courses will be sat in 1972. It is just as important to have well trained, high calibre technicians as technologists in industry. If this does not happen then many excellent ideas may exist with no means of putting them into practice!

The National Diploma in Agricultural Engineering was discontinued in the early 1970s. It is planned to set up another route to higher technician level at about the same time in the form of a Higher National Certificate in Agriculture which may be followed by a further course leading to a Farm Mechanisation endorsement. The effect of this endorsement is to bring the qualification into the Agricultural Engineering sector as shown in the diagram.

When I was making my way towards agricultural engineering, many of the routes as indicated in the diagram were not then defined; and the B.Sc. courses in agricultural engineering did not exist. The pattern of education has developed rapidly over the past few years in the applied sciences and perhaps most of all in agricultural engineering. We can expect further developments to take place which will be necessary to match the requirements of this rapidly growing subject.

I do not want to go into the details of the courses mentioned in the diagram here. The addresses given in the appendix indicate where up-to-date information can be obtained. I think my own sights would be set on a B.Sc. in agricultural engineering, but of course the particular route chosen would depend very much on circumstances and all the other factors that I have previously mentioned.

One of these factors was opportunity and what lies ahead in terms of career prospects. Diagram 3 opposite indicates the

IF I COULD START AGAIN 129

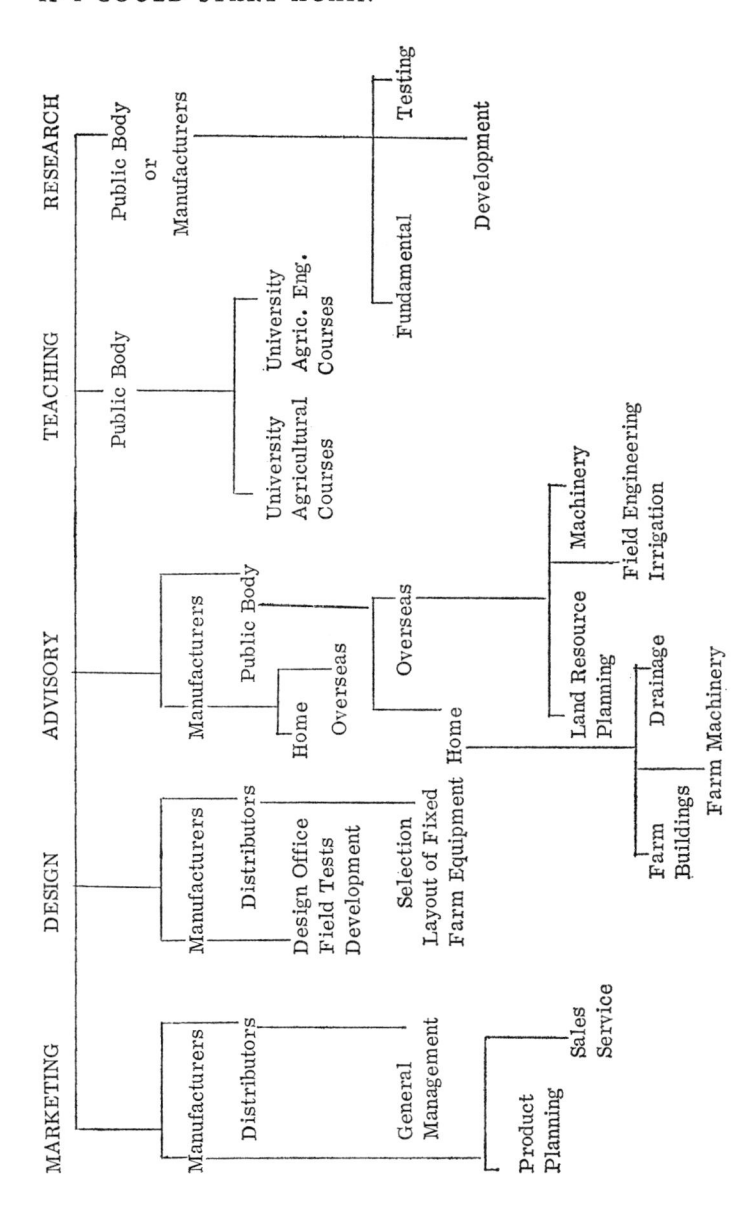

MARKETING
- Manufacturers
- Distributors
- General Management
- Product Planning
- Sales
- Service

DESIGN
- Manufacturers
- Distributors
  - Design Office
  - Field Tests
  - Development
  - Selection
  - Layout of Fixed Farm Equipment
  - Home

ADVISORY
- Manufacturers
- Public Body
  - Home
  - Overseas
  - Overseas
    - Land Resource Planning
      - Machinery
      - Field Engineering
      - Irrigation
    - Drainage
    - Farm Buildings
    - Farm Machinery

TEACHING
- Public Body
  - University Agricultural Courses
  - University Agric. Eng. Courses

RESEARCH
- Public Body or Manufacturers
  - Fundamental
    - Testing
    - Development

Diagram 3. Career Possibilities for Graduates in Agricultural Engineering types of job available to graduates in agricultural engineering. Our records of graduates' progress since leaving Silsoe match closely with these prospects. The few exceptions are mainly those graduates who have decided to use their broad training gained from a study of agricultural engineering as a springboard to other professions. These include Accountancy, Estate Management and Advertising!

Nowadays there are at least three recognised branches of agricultural engineering. This fact would present me with a difficult decision once I had completed my general education and training. All three branches are very much concerned with helping to provide more food as efficiently as possible in all parts of the world. But each branch has its own special way of contribution towards this end.

The first is farm machinery design and development. This is the area in which I have specialised after leaving College and would be a hot favourite again, I am sure. It seems to me that there is still so much to be done in this field. Many mechanical aids for farming already exist but more are required both in developed and developing areas of the world. More work is needed, for example, on the design and development of complicated machines like grain harvesters with automatic devices for controlling speed and perhaps direction of the machine so that it gathers the crop as quickly and economically as possible without damage. Perhaps these machines should be replaced altogether and an entirely new method of gathering the harvest devised.

The designer must try to decide what type of machines are required in the future. If man is to remain on the machine as an operator then designers have considerable responsibility to ensure that the operator can give his best performance without adverse effects on his health. The small tractor and equipment project which has been mentioned is only a part of the earliest technical programmes in developing countries.

There is an enormous challenge for design engineers working on mechanisation problems in the tropics. Farm machinery designers of the future have a considerable reputation upon which to build. Britain is the world's largest exporter of agricultural tractors and has been responsible for the introduction of many new ideas and improvements in mechanised farming. But there is much which remains to be done. If I were starting again then this would be the sort of challenge that I would find difficult to resist.

Innovation and development in agricultural machinery is important for increased food production but there are other equally important problems to be overcome, many of which may be found under the title of Field Engineering. If Mechanical Engineering is the closest relative to Farm Machine Design then Civil Engineering is next of kin for Field Engineering. I have already described the interest that I found in this work at Writtle and had I have been able to split myself in two at that stage, the other half would almost certainly have specialised in Field Engineering!

Before the earth or soil can support plant growth it must contain both air and water in just the right quantities. If there is not enough water then irrigation is required, but drainage will be necessary if there is too much. While the Civil Engineer might be responsible for bringing water to an area which is to grow crops it is the agricultural engineer who must get it from the reservoir or canal to the plant. To do this he has to use his knowledge of engineering and plant requirements.

A broad understanding of engineering and agriculture is also required for the solution of drainage problems. There are several dramatic examples in the world of failure to take account of soil erosion problems. Perhaps the best known is the Great Dust Bowl in North America. If agricultural engineers with a special interest in field engineering had been listened to at the time, the Dust Bowl problem might never have existed.

Many people in several countries are showing an increasing interest in the control of environmental conditions for both plants, animals and stored produce. The agricultural engineer is very much involved in the work. Those early tomatoes in the garden greenhouse would never have matured without their special environment! Other specialists have shown that temperature, light spectrum, humidity and carbon dioxide content in the air are all very important to plant growth. The agricultural engineer is now working on methods of controlling the plant environment to give the right amounts of these quantities. Animals also benefit from controlled environment and agricultural engineers are involved in the selection and use of materials for buildings which give the right conditions.

Within the next ten years new branches of agricultural engineering will be formed. The food processing industry is having an increasing influence on our daily menus. Prepacked and frozen foods are no longer uncommon. Synthetic foods have arrived on the table in the United States. It is only a question of time before they become a part of our diet in Britain.

The farmer will need to modify and change his methods to meet these new developments. The agricultural engineer must also ensure that he is always ready to contribute to and influence progress.

Having given myself the opportunity of starting again, I seem to have reached an impossible situation. Choosing agricultural engineering once more wasn't difficult but I now find that the subject has grown so much in scope, interest and challenge. It is going to grow much more in the future. What can I do now? Perhaps my three young sons will solve the problem for me by taking one of the three existing specialisations each. I shall have to wait and see!

APPENDIX:
SOME HELPFUL INFORMATION

IF YOUR INTEREST has been aroused while reading this book then you may wish to find out more information and details. The following addresses and notes will provide a starting point.

1. **Agricultural Engineering Education**

    Courses at technologist or degree level are provided by:

National College of Agricultural Engineering, Silsoe, Bedford.	Three and four year B.Sc. degrees. Higher degrees and postgraduate diploma studies.
University of Newcastle-upon-Tyne, 6, Kensington Terrace, Newcastle-upon-Tyne 2.	Three and four year B.Sc. degrees. Some higher degree work.
University of Reading, Reading, Berks.	Postgraduate degree level studies only.

    At higher technician level, courses are available at:

Essex Institute of Agriculture, Writtle, Chelmsford, Essex.	Higher National Diploma in Agriculture followed by Mechanization Endorsement studies.
The West of Scotland Agricultural College, 6, Blythswood Square, Glasgow, C2.	College Diploma in Agricultural Engineering.

Information concerning the recently introduced City and Guilds certificate courses can be obtained from the City and Guilds of London Institute, 76, Portland Place, London W1. Booklets outlining the Regulations and syllabuses for the examinations are available from the Institute. The 465 course can lead to graduate membership of the Institution of Agricultural Engineers. This course is currently being offered by Rycotewood College, Thame, Oxon, which also provides a course leading to the award of an Ordinary National Diploma in Engineering with an agricultural engineering bias. The City and Guilds of London Institute will also provide a list of other establishments offering the 465 Agricultural Engineering Technician's Certificate course.

2. Practical experience and training

This work is an essential part of career development in agricultural engineering. Preferably, the experience and training should be gained on mechanised farms, with agricultural machinery dealers or with manufacturers. It is impossible to give addresses of all these organisations but if you wish to spend some time on practical work during vacations or for a period between school and College then the following organisations would be able to tell you about the firms providing training or vacation employment.

Agricultural Machinery and Tractor Dealers Association, Penn Place, Rickmansworth, Herts.	A body representing agricultural machinery dealers in Britain.
Agricultural Engineers Association, 6, Buckingham Gate, London, S.W.1.	The organisation with which most agricultural machinery manufacturers are associated.

Farmers in your area would be likely to welcome additional assistance especially during harvest time!

3. Publications

There are of course many agricultural engineering textbooks available which cover various aspects of the subject at different levels. The following

APPENDIX 135

selection of journals and magazines will however, provide an up-to-date indication of current developments and trends in and around the profession.

The Agricultural Engineer
The Journal and Proceedings of the Institution of Agricultural Engineers published quarterly for the Institution of Agricultural Engineers, Penn Place, Rickmansworth, Herts.

Power Farming
A monthly magazine mainly intended for 'Progressive arable and livestock farmers' which contains several articles on mechanization developments and techniques. Published by Agricultural Press Ltd., 161–166 Fleet Street, London, EC4P 4AA.

Agricultural Machinery Journal
The journal of the agricultural and garden machinery industry also published monthly by Agricultural Press Ltd. New machinery, modifications to existing equipment and review articles feature prominently in this journal.

Span
Published quarterly by Shell International Chemical Company Ltd., Shell Centre, London, S.E.1. A 'House Journal' which presents a wide range of articles written by experts outside the Shell organisation. Several articles are relevant to agricultural engineering.

World Crops
This magazine deals with mechanization and agricultural problems in all parts of the world, particularly the developing countries. Published bi-monthly for Morgan-Grampian (Publishers) Ltd., 28 Essex Street, Strand, London, W.C.2.

Journal of Agricultural Engineering Research
Published quarterly for the British Society for Research in Agricultural Engineering. The Journal contains papers describing fundamental research in agricultural engineering.

4. Professional institutions

General enquiries about the profession should be addressed to the Secretary, Institution of Agricultural Engineers, Penn Place, Rickmansworth, Herts.

Addresses of other professional institutions may be obtained from your local Careers Officer or the Library.

5. Demonstrations and exhibitions

Equipment, machinery and buildings may be seen at the Royal Show which is normally held in early July at Stoneleigh in Warwickshire. There is also an annual exhibition of machinery and livestock at the Smithfield Show which takes place normally in early December at Earls Court, London.